"If you are an educator and want to efficiently teach collaborative mathematics to your class, this is the book for you! It is a fun, challenging, and playful way to introduce problem-based learning by providing all the tools and problems necessary to get started."

Michaela Hlasek,
Math Teacher and Combinatorics Instructor,
Awesome Math Summer Program 2017

"Awesome Math, appears to be about math, but really has lessons for education in general and even for re-skilling in the corporate world—an effective approach to educate and prepare the next generation for a YouTube + machine learning world. It gives new meaning to the phrase 'the journey is the reward.' The biggest danger? This book could convince you that math can be fun."

Raj Varadarajan,
Senior Partner and Managing Director, Boston Consulting Group

"This book is a brilliant road map that delights in its own theorem of authenticity and relevance. Full of the philosophical groundwork, expert insights, and plenty of practice problems, *Awesome Math: Teaching Mathematics with Problem-Based Learning* is a must-read for any Math or STEM educator concerned with the relevance and joy of a beautiful and expansive discipline."

Ben Koch,
Co-founder and CEO, Numinds Enrichment

"*Awesome Math* makes a strong case for ditching rote memorization and turning to collaborative problem-solving and mastery-based learning instead. This book is a must-read for parents and educators in all subject areas who wish to develop their students' creative and critical thinking skills."

Jaime Smith,
Founder and CEO of OnlineG3.com

"The book is an excellent source for educators interested in problem-based learning through student-centric approach. Students and teachers will find some 'secrets' of how math circles, math competitions, experiences of other math educators, and even math games along with wonderful and challenging problems can be used for an entire lesson or just as a mini-unit."

Dimitar Grantcharov,
Professor at University of Texas at Arlington

"Through playful problem solving, mastery learning, the three C's, and more, *Awesome Math* challenges the idea of a traditional, teacher-centric classroom. Kathy, Alina, and Titu are visionaries in the field of math education, and their book has sparked new inspiration for strategies that I am eager to utilize in my own math classroom."

Hannah Keener

"*Awesome Math* emphasizes the importance of collaborative problem solving in a classroom setting, featuring interesting and carefully chosen concepts and problems that can be used in a regular classroom and enrichment academic mathematics programs such as math circles or summer camps."

Zvezdelina Stankova,
Teaching Professor of Mathematics at University of California at Berkeley

"This inclusive book speaks in voices of the many. It has the irresistible flow of a well-curated social feed. There are shiny treasures to repost, 'today-I-learned' surprises to ponder, wise checklists to save, heartfelt polemics to debate—and so many kind math friends to meet!"

Dr. Maria Droujkova,
Founding Director of Natural Math

"I believe the most important goal of education is acquiring the ability to learn on your own. This book is mainly aimed at this goal and will help teachers and students improve their logical thinking, making them more independent learners and scholars."

Dr. Krassimir Penev,
Bergen County Academies

Awesome Math

Teaching Mathematics with Problem-Based Learning

TITU ANDREESCU
KATHY CORDEIRO
ALINA ANDREESCU

JOSSEY-BASS
A Wiley Brand

Copyright © 2020 by John Wiley & Sons, Inc. All rights reserved.

Published by Jossey-Bass

A Wiley Brand

111 River St, Hoboken, NJ 07030

www.josseybass.com

No part of this publication may be reproduced, stored in a retrieval system, or transmitted in any form or by any means, electronic, mechanical, photocopying, recording, scanning, or otherwise, except as permitted under Section 107 or 108 of the 1976 United States Copyright Act, without either the prior written permission of the publisher, or authorization through payment of the appropriate per-copy fee to the Copyright Clearance Center, Inc., 222 Rosewood Drive, Danvers, MA 01923, 978-750-8400, fax 978-646-8600, or on the Web at www.copyright.com. Requests to the publisher for permission should be addressed to the Permissions Department, John Wiley & Sons, Inc., 111 River Street, Hoboken, NJ 07030, 201-748-6011, fax 201-748-6008, or online at www.wiley.com/go/permissions.

Limit of Liability/Disclaimer of Warranty: While the publisher and author have used their best efforts in preparing this book, they make no representations or warranties with respect to the accuracy or completeness of the contents of this book and specifically disclaim any implied warranties of merchantability or fitness for a particular purpose. No warranty may be created or extended by sales representatives or written sales materials. The advice and strategies contained herein may not be suitable for your situation. You should consult with a professional where appropriate. Neither the publisher nor author shall be liable for any loss of profit or any other commercial damages, including but not limited to special, incidental, consequential, or other damages. Readers should be aware that Internet websites offered as citations and/or sources for further information may have changed or disappeared between the time this was written and when it is read.

Jossey-Bass books and products are available through most bookstores. To contact Jossey-Bass directly call our Customer Care Department within the United States at 800-956-7739, outside the United States at 317-572-3986, or fax 317-572-4002.

Wiley also publishes its books in a variety of electronic formats and by print-on-demand. For more information about Wiley products, visit www.wiley.com.

Library of Congress Cataloging-in-Publication Data is Available:

9781119575733 (paperback)

9781119575719 (ePDF)

9781119575702 (ePub)

Cover image: Wiley

Cover design: © Nadzeya_Dzivakova/iStock.com

Printed in the United States of America

FIRST EDITION

PB Printing V10015442_110719

*To our awesome community of colleagues, family,
and friends who inspire us daily and made this publication possible.*

Contents

Acknowledgments xi
About the Authors xiii
Introduction xvii

I. Why Problem Solving?

Chapter 1: Rewards for Problem-Based Approach: Range, Rigor, and Resilience 5
Range Ignites Curiosity 5
Rigor Taps Critical Thinking 9
Resilience Is Born Through Creativity 10

Chapter 2: Maximize Learning: Relevance, Authenticity, and Usefulness 13
Student Relevance 13
Mathematical Relevance 14
Mathematical Relevance: The Math Circle Example 16
Curriculum Relevance 18
Authenticity: The Cargo Cult Science Trap 21
Authenticity in Learning 22
Usefulness 25

Chapter 3: Creating a Math Learning Environment 27
Know Yourself: Ego and Grace 27
Know Your Students 30
Know Your Approach 35

Chapter 4: What Is the Telos? 47
Autonomy to Solve Your Problems 47
Mastery Through Inquiry 48
Purpose with Competitions 50
Quadrants of Success 52

Chapter 5: Gains and Pains with a Problem-Based Curriculum 57
Teachers 58
Students 61
Parents 67

II. Teaching Problem Solving

Chapter 6: Five Steps to Problem-Based Learning 75
Start with Meaningful Problems 75
Utilize Teacher Resources 79
Provide an Active Learning Environment 91
Understand the Value of Mistakes 97
Recognize That *Everyone* Is Good at Math 99

Chapter 7: The Three Cs: Competitions, Collaboration, Community 103
Competitions 103
Collaboration 107
Community 117
Aspire to Inspire: Stories from Awesome Educators 121

Chapter 8: Mini-Units 147
Relate/Reflect/Revise Questions 147
Roman Numeral Problems 148
Cryptarithmetic 151
Squaring Numbers: Mental Mathematics 155
The Number of Elements of a Finite Set 157
Magic Squares 159
Toothpicks Math 163
Pick's Theorem 165
Equilateral versus Equiangular 168
Math and Chess 170
Area and Volume of a Sphere 172

III. Full Units

Chapter 9: Angles and Triangles 177
Learning Objectives 177
Definitions 177
Angles and Parallel Lines 177
Summary 180

Chapter 10: Consecutive Numbers 185
Learning Objectives 185
Definitions 185

Chapter 11: Factorials! 191
Learning Objectives 191
Definitions 191

Chapter 12: Triangular Numbers 199
Learning Objectives 199
Definitions 199

Chapter 13: Polygonal Numbers 205
Learning Objectives 205
Definitions 205

Chapter 14: Pythagorean Theorem Revisited 213
Learning Objectives 213
Definitions 213
Pythagorean Theorem 214
Rectangular Boxes 214
Euler Bricks 216
Assessment Problems 219

Chapter 15: Sequences 221
Learning Objectives 221
Definitions 221
Introduce a Geometric Progression 222

Chapter 16: Pigeonhole Principle 227
Learning Objectives 227
Definitions 227

Chapter 17: Viviani's Theorems 235
Learning Objectives 235
Definition 235

Chapter 18: Dissection Time...239
Learning Objectives 239
Definitions 239

Chapter 19: Pascal's Triangle...245
Learning Objective 245
Summary 249

Chapter 20: Nice Numbers..255
Learning Objectives 255
Definitions 255

Index...259

Acknowledgments

Special thanks to Navid Safaei and Alessandro Ventullo for their time in reviewing the mathematical content for this book.

– Titu and Alina

My heartfelt appreciation goes to my closest community, my family, for their support, advice, and contributions to this effort. To my husband, David, whose ideas and insights have added value not only to this book, but to our family for over 25 years. To my oldest son, Jacob, for his incredible gift of explaining complex concepts elegantly and easily, which helped improve sections of the book. To my youngest son, Adam, for his content corrections and positive support that kept me on track and enjoying the process. And lastly, to my mother-in-law, Sandy, and my sister, Kelly, for being early readers and emotional support. Thank you.

– Kathy

We'd like to thank Amy Fandrei, our executive editor, for her kind guidance and for providing us with the opportunity to share our love of problem-based learning. Many thanks also to Pete Gaughan, the content enablement manager for this project, who helped us every step of the way to create a quality publication.

– Titu, Kathy, and Alina

About the Authors

Dr. Titu Andreescu has been coaching, teaching, and training students and teachers for most of his exemplary career. Starting as a high school mathematics teacher in Romania and later in the United States, Titu became coach and leader of the United States International Mathematics Olympiad team, director of the Mathematical Association of America's AMC tests (American Mathematics Competitions), and an associate professor at University of Texas at Dallas in the Science and Mathematics Education department training mathematics teachers. His passion for problem solving and mathematics teaching has extended to the following noteworthy accomplishments.

- **AwesomeMath Summer Program** is a premier mathematics camp held on the campuses of the University of Texas at Dallas, Cornell University, and the University of Puget Sound. Awesomemath.org
- **AwesomeMath Academy** provides enrichment opportunities for students seeking a strong problem-solving–based curriculum with classes offered in North Texas and online. AwesomeMathacademy.org
- **AwesomeMath Year-Round** is a correspondence-based program that provides students with further opportunities to broaden their mathematical horizons, particularly in those fields from where Olympiad problems are drawn. https://www.awesomemath.org/year-round-program/
- **XYZ Press** (separate business entity affiliated with AwesomeMath) is the publication company that was started in 2008 to more efficiently bring problem-solving books to market. https://www.awesomemath.org/shop/about-xyz-press/
- **Mathematical Reflections** is a free online journal aimed at high school students, undergraduates, and everyone interested in mathematics. https://www.awesomemath.org/mathematical-reflections/

Purple Comet! Math Meet has been "fun and free since 2003." This annual, international, online, team mathematics competition is designed for middle and high school students. http://purplecomet.org/

Metroplex Math Circle is a free program that was designed to attract gifted students and educators in the Dallas/Fort Worth area to provide an avenue outside the standard curriculum to develop their mathematical and problem-solving skills. Further, the circle offers access to math competitions for students (in 2017–2018 school year, approximately 150 students participated in the AMC 8, 10, 12, and AIME competitions) each year who may not be able to participate in their schools. Metroplexmathcircle.org.

The **Math Rocks** curriculum, developed by Dr. Andreescu in 2008–2010, is still going strong in the Plano, Texas, school district for elementary and middle school students. The success of the curriculum has resulted in its extension to over 45 public elementary and 15 middle schools. http://k-12.pisd.edu/currinst/elemen/math/MathRocksInformation.pdf.

For **Kathy Cordeiro**, innovation, problem solving, and team collaboration have been the leading constants throughout her varied career. A degree in communications, coupled with an MBA, has given Kathy a unique skill set to create and market customized education initiatives, in business and/or academia, which allows her customers and students to reach their goals and realize success. Kathy began her own enrichment school, Eudaimonia Academy (2006–2012), where she coached math teams, taught a philosophy/creative writing course, and co-led speech and debate teams.

Kathy is the marketing and communications director for the AwesomeMath organization. In this role, she has had various speaking engagements as well as managed multiple communication channels online, where she discusses mathematics education with parents, teachers, students, and businesses.

Beyond being connected with multiple math groups, Kathy is also a part of a network that includes parents, teachers, and students, such as

- AwesomeMath parents, students, alumni
- Purple Comet supervisors/teachers
- Davidson Young Scholars, parents, and alumni
- Mathematics organizations
- Homeschool groups

Alina Andreescu was born and raised in Romania, at a time when mathematics education was exceptionally strong. She participated successfully in Romanian mathematics competitions. She completed her finance degree in the United States and later obtained an M.A. in management with emphasis on leadership. Alina was never afraid of change and challenges, embarking on lifetime journeys from moving to the United States to becoming a successful cofounder and leader of the AwesomeMath and XYZ Press organizations.

As operations director of the AwesomeMath programs for the past 12 years, Alina has been integral in every facet of creating the opportunities/resources that fulfill the mission of providing enriching experiences in mathematics for intellectually curious learners. She fosters a community of staff, students, and instructors that values critical thinking, creativity, passionate problem solving, and lifetime mathematical learning. Since the AwesomeMath community is international, she must meld a diverse background of individuals into a thriving learning environment.

Introduction

In writing this book, we hope to lead you to what you already know: that problem-based learning is an effective method for raising tomorrow's thinkers by collaborating over interesting and relevant problems. Through the AwesomeMath Summer Program,[1] the inspiration for this book, we've had the privilege to work with thousands of the brightest minds from around the globe for over 10 years. We've seen first-hand the leaps in skills, growth of curiosity, and joy of problem solving that arises when individuals are immersed in a kind, collaborative, and challenging environment where students create positive life-long memories and form valuable friendships.

So, how do you raise out-of-the-box thinkers in a check-the-box world? Teaching is an opportunity to inspire and guide, but that means diverging from the conformity required in today's education system and allowing students to take intellectual risks and, yes, fail. The outdated criterion of identifying top students through grades is flawed; it's evaluating someone's worth based on an outcome and not the process, which sets up situations where students avoid intellectual risks so they can maximize grades. Students aren't learning how to *think, work together,* or *find challenging opportunities.*

Furthermore, they aren't being prepared to face the current challenges in today's workforce, which values innovation, leadership, collaboration, resilience, and critical thinking. We need students who can do more than solve mere exercises for a check mark; they need to be able to tackle difficult problems and also be able to notice problems worthy of solving by seeking patterns, reframing information, and asking the right questions. Students are all different and have different strengths to offer in every setting. We need to value them for who they are with a student-centric approach as opposed to evaluating them with standardized conformity and false metrics.

When Randy Pausch gave his Last Lecture at Carnegie Mellon University,[2] he explained that the moment someone lowers their expectations for what you can accomplish, they've stopped caring about you. Students need to have challenges that we, as educators, know they can overcome and master. When we allow students to work toward mastery instead of grades, then the journey becomes about the process and not the outcome. This approach, however, requires facilitators, helpers, and guides along the way so that each student can recognize their value and be their best version.

Problem-based learning approaches education with a deep respect for the value, abilities, and strengths of each student by raising expectations beyond the standard and providing guidance in a supportive environment.

The main goals of this book are as follows:

- To show that a problem-based curriculum is an effective way to teach mathematics to students of all levels and backgrounds and prepares them to be creative thinkers in an ever-changing world.
- To train educators on how to employ a problem-based curriculum in their classrooms by creating a collaborative, kind, and engaging environment where each student can be guided to be their best version.
- To provide the curriculum plans and interesting problems that allow educators to successfully train their students to think with a problem-solving mindset.

Here are the top five characteristics of a problem-based learning curriculum as detailed in this book:

1. It is *student-centric* as opposed to teacher-centric. Lectures are kept as brief as possible and students are the drivers in the process while teachers are the facilitators of learning.
2. It is *highly collaborative* because when you engage in the trade of ideas, everyone improves.
3. It is *scalable* so that problems are in a range to reach all levels of students and promote their individualized growth.
4. It relies heavily on *range, rigor, and resilience* to encourage curiosity, critical thinking, and creativity.
5. It is *FUN*! If the teacher and students have the correct mindset of playful mathematics and growth in a supportive environment, then they look forward to the lessons and don't resist extra challenge.

A typical adult gorilla is 5 ft tall and weighs 400 pounds. If King Kong is 20 ft tall, how much does he weigh approximately? This problem is about basic measurements; however, many students get it wrong by rushing to provide an answer without *thinking* about what really is being asked. Many will quickly respond, "1,600 pounds," which is completely illogical if they have a sense of weight.

Only 1,600 pounds for such an enormous gorilla? A typical black and white cow weighs that much! A hippopotamus can weigh up to 4,000 pounds. While this is a simple exercise rather than a complicated problem, it illustrates a larger problem in mathematical thinking.

When mathematics pedagogy is reduced to checking a box, guessing an answer, or completing repetitive exercises, then students are rewarded for quickly reaching a solution over thoughtfully working through a problem.

Solution
You know the height difference (one dimension), but you need to translate that to the difference in volume (three dimensions) of the gorillas. If King Kong is four times bigger in each of the three dimensions (4 times taller, 4 times wider, and 4 times longer) than the average gorilla, that equals 400 multiplied by 4 multiplied by 4 multiplied by 4, i.e., $400 \times (20/5)^3 = 25\,600$ pounds.

Notes
1. https://www.awesomemath.org/what-is-awesomemath
2. Randy Pausch, *Last Lecture: Achieving Your Childhood Dreams*, Carnegie Mellon University, December 20, 2007, https://www.youtube.com/watch?v=ji5_MqicxSo.

SECTION I

Why Problem Solving?

In this section:

- Rewards for a Problem-Based Approach: Range, Rigor, and Resilience
- Maximize Learning: Relevance, Authenticity, and Usefulness
- Creating a Math Learning Environment
- What Is the Telos?
- Gains and Pains with a Problem-Based Curriculum

Today's kids are busier than ever! Juggling their schedules inside and outside of school requires major planning, and as a result, enticing them to focus in a mathematics class can be difficult. That is not to say that they are incapable of deep thought, but rather, asks how mathematics can compete with all the other distractions that life throws their way. What makes activities such as sports or video games so much more appealing? How can we construct a mathematics environment so that students are engaged with the subject and work together to achieve a superior understanding for mathematics?

The common thread is playful problem solving. *Play* is an integral part of life. Even as adults, we love to play and compete and solve problems with friends. You can challenge yourself to move up levels and share your experiences with peers – plus, there is no fear of losing, whereas in mathematics, there is fear. Fear of appearing stupid, fear that if you are slow to understand that you just aren't *good at math,* fear that doing poorly in math means you won't get into college. We need to erase that fear and help kids take thought risks with problem solving.

Let's say that Steve is playing a video game with a friend and loses a boss battle. Will he give up and say, "Well, I guess I'm bad at video games, so I'll stop playing"? Of course not. He'll try a different strategy or ask his friend for advice or go online and watch YouTube videos. What makes the difference between perseverance and giving up? Mathematics education *can* be just as playful and allow students to compete by solving meaningful problems while working as a team, but that means the stakes need to change, and instead of teachers as judges, saying a student's individual work is good or bad, is missing steps, is B work and not A work, they need to shift into being coaches who guide their students to being the best versions of themselves.

While teachers want each student to excel, in reality, great teachers work on improving the abilities of their entire class every day, spotting areas that are weak, celebrating strengths, and being a cohesive unit. When all of those areas come together, then success will happen. *Children are not outcomes* and need to be guided by a great educator to think critically and creatively.

Currently, math education in middle and high schools is a series of exercises with easily obtained answers, e.g., find the perimeter of a square, training students to do what a computer can do better. Problem solving goes much deeper and taps into what makes us human, namely, multiple creative approaches with a string of steps to solving meaningful and interesting problems. It takes the shift away from *outcome*-based learning (grades/test scores, rank, grade point average [GPA]), which is a fixed-mindset approach, to learning for *mastery*, where students challenge themselves to improve every day (growth mindset).

What exactly *is* problem solving? Even mathematicians and researchers haven't come up with a definitive answer, but in this book, we believe problem solving has the following characteristics:

- Problems take several steps to solve.
- More than one approach can be used to arrive at a complete solution.
- Good problems lend themselves well to collaboration with peers.
- Meaningful problem solving promotes flexibility of thought and innovation.
- Mathematical learning and reasoning are integral to the process of problem solving.
- Problem solving is about working around obstacles to understand the unknown.

Problem solving is the strategy, and math competitions are the vehicle to train your math class to be stellar thinkers. Since the current school curriculum delivers a narrow path of mathematics knowledge, climbing aboard the math competition train will expose students to a greater array of topics, including *discrete mathematics*, an area that incorporates both number theory and combinatorics (counting and probability). Discrete math, along with finite mathematics and linear algebra, are necessary to work in the modern world of computing. Mathematical modeling and a strong understanding of statistics is also critical. The level of deep thinking required to solve hard problems in the areas of discrete mathematics, algebra, geometry, and the areas in between (e.g., geometric inequalities), transfers to future careers in STEM (science, technology, engineering, mathematics) fields, and beyond. Mathematics competitions provide exposure to all these topics while working with peers to solve challenging problems.

Just as every football player cannot be the quarterback, not every student is going to excel in the same way with mathematics competitions, but this brings us back to the focus being placed on the process and not the outcomes. The reason to engage in math competitions is to have something to work toward where each student can get a little better every day and be motivated in a collaborative and supportive environment. Some students may enjoy working through lots of different types of problems while others may prefer to look at the methods employed and want to write their own problems based on their discoveries. Every type of student can play an important role in your mathematics class, and as the teacher, you want to look at every student as a collection of strengths as opposed to a collection of weaknesses that need to be fixed.

Regardless of the role a student chooses, all students grow their skills faster when collaborating toward a common goal than they would on their own, because when you engage in the trade of ideas, everyone improves.

The learning environment for the game is critical to bringing out the best in the players and the rewards are *range*, *rigor*, and *resilience*.

What Were Your School Experiences Like in Your Country That Contributed to Your Love of Problem Solving?

Dr. Branislav Kisačanin: When I was growing up in former Yugoslavia, during the 1980s, math and physics competitions were well organized and students were encouraged to participate. Competitions were held at school, city, regional, and national level, and from there teams were sent to the International Mathematics Olympiad (IMO) and the International Physics Olympiad (IPhO). Except for the school level, all competitions involved some kind of travel with like-minded kids, and that was a big part of it all for me. Thanks to these competitions, I met many life-long friends (my fellow students and my future college professors) and visited wonderful places in former Yugoslavia: Postojna cave and Portoroz in Slovenia, Sarajevo in Bosnia, Decani, with its famous fourteenth-century monastery, and the Danube's Djerdap Gorge in Serbia.

CHAPTER 1

Rewards for Problem-Based Approach: Range, Rigor, and Resilience

Range Ignites Curiosity

As educators, we understand the importance of depth and breadth in learning. For beginning piano players, listening to a concert pianist perform can ignite curiosity and inspire them to practice more. In mathematics, there seems to be a reticence to hear the symphony for fear that it will be too much, too soon, and by limiting the range, we limit curiosity and growth.

Our AwesomeMath Enrichment programs are filled with students whose schools placed a ceiling on their mathematics education and they are seeking outside resources. Sometimes, this is because the school is concerned that if students are accelerated too quickly, they won't have the maturity to truly understand what is being taught. Other times, the school is concerned that if students move two or three levels ahead, by the time they are seniors they will have run out of classes to take. Another common reason our parents provide is that the school thinks it will be *too much* information for the student at too young an age and will spoil their performance in future classes. With a problem-based curriculum, there is no ceiling on learning and there is ample depth and breadth of subjects to keep students challenged throughout their lifetime. The real danger of not giving students adequate challenge and range to satiate their curiosity is that they will turn off on mathematics and learning altogether. Students need to hear the beauty and art that is mathematics to kindle their joy of learning.

> ### Why Is a Collaborative Problem-Based Approach Worthwhile?
>
>
>
> *Dr. Emily Herzig:* Collaboration in the classroom has many benefits. Research has demonstrated that active learning improves performance on exams, and the effect is especially large for disadvantaged students. Currently, education is not equitably accessible to all students, with students from underserved populations and first-generation college students in particular facing additional obstacles to entering, navigating, and excelling in higher education. Thus, collaborative learning in the classroom could be key to closing the achievement gap and allowing capable but underprepared students to reach greater success in math.
>
> Furthermore, a collaborative and problem-based approach gives younger students a more accurate impression of what higher-level math entails. Students too often carry the belief that success in math is based in rote memorization and drilling problems. While those skills are certainly useful for efficiently carrying out the basic mechanics of solving problems, it is equally important that students are able to formulate and interpret more complex problems, and work with their colleagues to develop and execute problem-solving strategies. Arguably, this process is also what makes math such an enticing subject. A focus on collaborative problem solving is a great way to attract students to and prepare them for careers in math.

Even the terms used for learning piano and learning mathematics are different: Students *play* piano and **work** on math problems. There needs to be a fundamental shift in approach and exposure to a range of problems that are harder and more interesting so that students can see where math can take them. So much of math education today is about *waiting*:

- Wait until high school, and then what you've been learning in middle school will be useful.
- Wait until college, and then what you've been learning in high school will be useful.
- Wait until you learn topic x before you can see the beauty of topic y or z or beyond.
- Wait until you learn a subject, like geometry, in isolation before you have the ability to learn how it connects and contributes to other areas such as algebra, engineering, art, science, etc.

And so on. . . .

A mathematician, like a painter or poet, is a maker of patterns. If his patterns are more permanent than theirs, it is because they are made with ideas.

G.H. Hardy

Students need to *work together* to weave a pattern of ideas with what they know and can add more as their knowledge base grows. To fully appreciate the full tapestry of mathematics is not by adding one color of thread at a time, but weaving a picture with various thread colors within a group of learners who are just as excited by the beauty as you are.

CHAPTER 1: REWARDS FOR PROBLEM-BASED APPROACH: RANGE, RIGOR, AND RESILIENCE

Collaborative problem-based learning holds the key to unlocking abilities and mathematical growth. The collaboration nurtures interpersonal skills, leadership traits, and conflict resolution, while providing interesting problems gives students a common goal to work toward where they can connect and share ideas. Students can see the fun and beauty in mathematics, start to play, and see where math can take them long term.

A fun problem that works well in a group is the following:

> The diagram shows a polygon made by removing six 2×2 squares from the sides of an 8×12 rectangle. Find the perimeter of this polygon.
>
> *The answer is 60.* The square removed from the lower-right corner of the rectangle does not change the perimeter of the polygon, but when each of the other five squares is removed, the perimeter is increased by 4. Thus, the perimeter of the polygon is $2 \times 8 + 2 \times 12 + 5 \times 4 = 16 + 24 + 20 = 60$.[1]

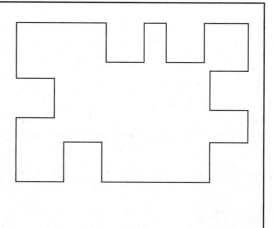

Students are allowed to be transported by a musical symphony before understanding individual notes, so why not provide the symphony of mathematics and introduce students to its wonders and challenges? Parents and educators will read books beyond a student's personal reading level so that they can hear the richness of language and be exposed to more intricate sentence structure. There are so many wonderful places an enriching math education can take you.

When educating young math students, you can let them know that they are the captains of their ship, but as their navigator, you can guide them to really interesting destinations and expose them to a wider range of mathematics.

 Following are some ideas for increasing range in a mathematics program:

Logic problems. Thinking through these problems primes the brain for mathematical reasoning.

Combinatorics. This discrete mathematics field involves counting and probability.

The mathematics of science. Solving, for example, real-world physics problems adds range and connectivity across academic disciplines.

Mind benders and puzzles. These are a fun way to introduce mathematical concepts, such as magic squares.

Game theory. Utilize mathematical modeling to understand the strategies employed by rational game players (decision makers).

Computational linguistics. Apply computational analysis to language and speech for linguistic phenomena.

And so much more!

There is a world of mathematics to discover, and the playful pursuit will ignite their interest and provide them with the introspection to know their strengths and passions. Context while explaining mathematical concepts is also important, so it's not just range within the topic of mathematics but range outside the topic as well. Knowing the history and story of what they are learning makes a huge difference. Mathematical discoveries were made to solve real-life problems, and if students learn the story, they are more connected with the material.

For example, students will often ask, "When will I ever really need to use algebra?" It's easy to give them the F.U.D. answer (a marketing term that means Fear, Uncertainty, and Doubt to nudge consumers into decisions): "If you don't learn algebra, you won't do well on the SAT, and then you won't get into a good college." That's a lousy answer, and unfortunately is the message heard by many students, either directly or indirectly.

When running a mathematics club, a student asked this exact question, even though her mother had majored in math when she was in college. In response, she was assigned a one- to two-page paper and oral report on the importance of algebra, which, as you can imagine, wasn't initially well received. The paper needed to include:

1. Research the story of algebra (why, how, and by whom was it created).
2. Describe where algebra is used in life (why, how, and when it is necessary today).
3. Explore how algebra would be helpful in her own life.

She was given the resources to get started, and then she knocked this assignment out of the ballpark. Her understanding deepened and her love for the topic, in turn, grew.

Below is an algebra problem that students can also use logic to figure out. These problems help engage the class and get them thinking!

> I have two hourglass clocks filled with sand. One empties in 4 minutes, the other in 7 minutes. How can I use them to measure exactly 10 minutes?
>
> **Solution:** Start both clocks at the same time. When the 4 minute timer runs out, start it again. When the 7 minute timer runs out, turn the 4 minute time over and it will run for the required extra 3 minutes.
> (*Math Leads for Mathletes*, Book 2, page 48)

Providing a *range* in topics and connecting mathematics with other disciplines, such as history, will ignite students' *curiosity* to dig deeper and find the beauty and relevance of the mathematics they learn.

CHAPTER 1: REWARDS FOR PROBLEM-BASED APPROACH: RANGE, RIGOR, AND RESILIENCE

Studying ancient number systems is a great way to show the progression of numbers from being an adjective, e.g., a one-to-one correspondence as in one sheep per tally mark, to being a noun, e.g., thinking of numbers in the abstract. Students are a microcosm of what took ancient civilizations thousands of years to understand.

These systems are also a great way to learn about base systems beyond base 10. Babylonians used a hexigesimal system (base 60) and the Mayans used a vigesimal (base 20) system, probably because in the warmer climate, they used their fingers and toes for counting. It's an easier mental leap to begin to learn binary systems (base 2) that are integral in today's digital world.

When exploring a range of topics, this is not something to be done alone. Exploration should, of course, be a shared experience. Students need to present what they've learned to the class so that they can all grow together and gain the confidence for bigger challenges.

Rigor Taps Critical Thinking

With the high-pressure stakes of standardized testing, rigor is starting to take on a negative connotation. It's been used to *pump and dump*, namely, shove as much information as possible into a student so that they, in turn, can dump that information on to a bubble sheet of multiple-choice answers.

In this book, we refer to rigor as a way to tap critical thinking so that students can have meaningful experiences and try novel approaches to solve problems. That means presenting the student with *problems* instead of *exercises*. Math education must be more than a series of easily obtained answers (exercises) (e.g., find the perimeter of a square, training students to do what a computer can do better). Problem solving goes much deeper and taps into what makes us human, namely multiple creative approaches with a string of steps to solving challenging and interesting problems.

Scaffolding problems based on difficulty adds rigor and progression for students of various skill levels so as to reach a wider range of kids. Below are some example problems from the book *Math Leads for Mathletes*, Book 2, centered on least common multiple (lcm). You can always start with exercises as a warm-up, but then ramp up the challenge to more thought-provoking problems.

PROBLEMS

1. What is the least common multiple of 6, 8, 24, and 30?
2. What is the least common multiple of 585 and 10 985?

3. Four cargo ships left a port at noon, January 2, 2010. The first ship returns to this port every 4 weeks, the second every 8 weeks, the third every 12 weeks, and the fourth every 16 weeks. When did all four ships meet again at the port?

4. It is given that the number of spots on a Dalmatian is less than 20. Also, the number of spots is divisible by 3. Furthermore, when the number of spots is divided by the number of legs, the remainder is 3. Finally, the number of spots leaves a remainder of 6 when divided by 9. Find the number of spots on the Dalmatian.

5. I am thinking of a number. The least common multiple of my number and 9 is 45. What could my number be?

SOLUTIONS

1. The lcm $(6, 8, 24, 30) = 2^{(3 \times 3 \times 5)} = 120$.
2. The lcm $(585, 10\,985) = 3^{(2 \times 5)} \times 13^3 = 98\,865$.
3. The lcm $(4, 8, 12, 16) = 2^{(4 \times 3)} = 48$. Hence, all ships will meet again at the port in 48 weeks' time, on December 4, 2010.
4. The lcm $(3, 4) = 12$. Adding the remainder 3 gives 15. We find lcm $(3, 9) = 9$. Adding the remainder 6 gives 15 as well. The lcm $(3, 4, 9) = 36$; therefore, other possible solutions are $15 + 36 = 51$, $15 + 2 \cdot 36 = 87$, $15 + 3 \cdot 36, \ldots$, but 15 is the only positive solution less than 20.
5. The prime factorization of $45 = 3^2 \cdot 5$. Therefore, the number could be 5, 15, 45 because lcm $(5, 9) =$ lcm $(15, 9) =$ lcm $(45, 9) = 45$.

Teaching through challenging problems requires rigor on both the part of the educator and the student, but also requires maintaining a balance between providing these problems and learning foundational concepts. If students can be exposed to the tough problems in mathematics early on, this rigorous training will make other challenges easier to face. Just as it's easier to jog a mile if you've been training for harder races, when you learn the critical thinking that comes along with problem solving, then every other topic that requires math (and most do, in one way or another, e.g., economics) will be easier to understand.

Rigor is both the result of critical thinking and creativity as well as the discipline to learn the foundational concepts necessary to be successful. This rigor is easier to manage when you have a collaborative team (the teacher and fellow students) offering support and encouragement along the way. This leads to the next reward: resilience.

Resilience Is Born Through Creativity

Making the simple complicated is commonplace; making the complicated simple, awesomely simple, that's creativity.

Charles Mingus

Problem solving requires creativity to reach a solution, as there isn't always just one clear-cut approach. Creativity requires bravery and resilience. People feel that creativity just happens, when in actuality it is the product of trial and error – requiring the resilience to persevere. Pushing through failures to the other side is a worthy goal because the benefits are so high, such as developing the resilience to take on harder and harder challenges.

In competitive chess, coaches will say you need to lose 10 000 games to become a grand master, so whenever a student loses a game or makes a mistake with a problem, *celebrate!* You are on the road to mastery, and the resilience of making mistakes will take you there.

Life is not a zero-sum game (in game theory, this is the mathematical representation that one player's gain is equivalent to another player's loss). Mistakes make us better and help to develop a growth mindset when viewed in the proper light.

Students, through rigor and range, grow their courage for trying new things as well as hone their creativity and resilience through constraints. How are creativity and constraints related? Math, in and of itself, can be a beautiful and creative pursuit – elegant proofs, creative problem solving, and being connected to a global community where ideas can flourish. That creativity, of course, operates within the constraints of mathematical reasoning and rigorous proofs.

There are myriad problems where you have to jump in headfirst and creatively play before a solution will present itself to you. They take time, patience, and resilience to continue and chip away at them, and that process can be the most exciting part.

Following is a problem that requires a lot of thought and needs a well-conceived plan of attack, building resilience while also providing the thrill of solving!

The diagram here shows a 12 by 20 rectangle, split into four stripes of equal widths, all surrounding an isosceles triangle. Find the area of the shaded region.

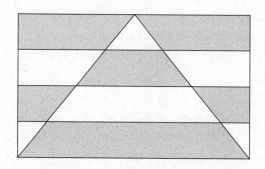

The answer is 150. The triangle has base 20 and altitude 12, so its area is $(20 \times 12)/2 = 120$. Each strip has a width that is ¼ the height of the entire triangle, so the small triangle in the top strip has area $120 \times (¼)^2 = 15/2$. Similarly, the area of the triangle within the second strip has area $120 \times (½)^2 - 120 \times (¼)^2 = 45/2$. The area of the triangle within the third strip has area $120 \times (¾)^2 - 120 \times (½)^2 = 75/2$, and the area of the triangle within the fourth strip has area $120 - 120 \times (¾)^2 = 105/2$. Each strip is ¼ the width of the entire rectangle, so each has area $¼ \times 12 \times 20 = 60$. It follows that the sum of the shaded areas within the four strips is $(60 - 15/2) + 45/2 + (60 - 75/2) + 105/2 = 150$.[2]

Because of these constraints, people will reach the conclusion that mathematics is dry and formulaic, but anyone that has ever delved deeper beyond the typical classroom approach to math understands how these constraints free the mind and lead to amazing discoveries. Certainly, math isn't the only area where constraints can be creative. One only has to listen to minimalistic music, read poetry, or see modern art for examples of limiting choices to create simple and elegant forms. Many interesting and delectable meals have been invented by people who are constrained by budget, choices, or geography.

Notes

1. Purple Comet Math Meet! 2019 contest, http://purplecomet.org.
2. Purple Comet! Math Meet contest 2019.

CHAPTER 2

Maximize Learning: Relevance, Authenticity, and Usefulness

Student Relevance

When mathematics is relevant to the students and their world, they will become more connected with the process and, in turn, recognize their own contributions and value. How can you maximize mathematical thinking in the classroom and increase that connection? How do you respond to the students who say that they "aren't good at math," as if being good at math just happens and the assessment is a binary choice?

Many times, the students will make this assessment of their math abilities based solely on the criteria of how *fast* they are at solving computations compared to their peers. Mathematics is *not* speed. Instead of worksheets with short exercises, timed tests, and/or reducing mathematics to arithmetic, students need to be exposed to meaningful and challenging problems where their personal strengths can shine. *When students are viewed as a collection of strengths rather than a collection of weaknesses that need to be fixed, they will see the value they can add working on a fun, new challenge!* This value will shine through even more when they collaborate with a group of people working together to solve a difficult problem.

If a student struggles with a certain concept or idea in the lesson, they need to think of the flip side of this struggle. Namely, when there is a flaw in thinking, it usually has a corresponding feature in another area. In programming, there is the old joke, "It's a feature, not a bug," but individuals are complicated creatures, and those bugs are only viewed with one type of lens.

When the lens changes from the *person to the problem*, students won't fixate on whether they are personally *good* or *bad* at math, and instead will seek out what is needed to rise to the challenge presented.

- Do they need more information to understand the problem?
- Do they see a pattern?

- A dyslexic student may have difficulty focusing on word problems but excel in three-dimensional thinking, geometry, and visual proofs.
- A student who processes information through movement, meaning that they have to fidget or move around the classroom to think, may benefit from mental math challenges that don't require paper, so they can get the wiggles out while learning in the process.
- A reflective thinker may need to see problems a day ahead so they can have the necessary time to process what they are reading and employ their powers of thinking deeply and slowly.

Athletic coaches can split their teams into groups to work on different training exercises, and so can mathematics teachers.

- Can they create a diagram, chart, or mathematical model to help?
- What type of reasoning/strategy is required for the problem?
- Can they think of a simpler example?

As the teacher, you can guide them on their journey of discovery and help give them the confidence and skills to attack any problem that is presented. This will allow students to see the *relevance* of mathematics and be able to attach that relevance to other problems in academics and life.

> *The essence of math is not to make simple things complicated, but to make complicated things simple.*
>
> S. Gudder

Mathematical Relevance

Why should math be relevant?

- It can bridge learning to other areas of academics and life.
- It can enhance a students' learning experience because math is never-ending, and they can always challenge themselves at any stage.
- It has interesting topics and problems that keep students curious and engaged.
- It connects students with a community of thinkers so mathematical knowledge can be shared and enjoyed.

A problem-based learning curriculum provides all these points for making mathematics relevant. By engaging students with thoughtful topics, you will capture their curiosity and inspire them to dig deeper.

CHAPTER 2: MAXIMIZE LEARNING: RELEVANCE, AUTHENTICITY, AND USEFULNESS

Providing challenging problems is critical for making math pertinent and thrilling. Math competitions can take this approach even further by:

- Uniting students toward a common purpose, which is succeeding in competitions and improving their scores.
- Providing a collaborative environment, which is the most effective way to work through complex problems.
- Exposing students to discrete mathematics, which is critical for living in the modern world of computing and data.
- Creating a team atmosphere, which helps students step outside their comfort zone and promotes enthusiasm and productivity when taking on more difficult challenges.
- Supplying teachers with a treasure trove of problems for students of all levels, which aids lesson planning and saves time.
- Inspiring students to challenge *themselves* by developing their competition mindset, further allowing them to improve their abilities and learn new techniques.

Math competitions tap into the human need for connection and purpose, but only if the environment focuses on the process and not the outcomes. It's not about winning; it's about individual improvement. This is what makes students want to participate; they see the personal relevance, but they also have a community dedicated to seeing them improve and excel. It's more *coopertition* than competition, a portmanteau of cooperation and competition, where students work together and celebrate each other's successes.

> *Solving a problem for which you know there's an answer is like climbing a mountain with a guide, along a trail someone else has laid. In mathematics, the truth is somewhere out there in a place no one knows, beyond all the beaten paths. And it's not always at the top of the mountain. It might be in a crack on the smoothest cliff or somewhere deep in the valley.*
> Yoko Ogawa, *The Housekeeper and the Professor*

That means setting the tone from the beginning that mathematics is not something to be feared and that *mistakes are positive* and necessary for growth. In today's culture, you can be a star athlete and take pride in your work and accomplishments, yet star mathletes need to be humble and keep their accomplishments hidden so that others don't feel *less than*. This is all wrong.

Students who are strong in math and train hard to learn and grow should be able to have the same pride and be a part of an environment that can offer additional challenges, namely scalable curriculum. Students who struggle more can not only be inspired, just like those who are inspired by sports athletes, but also find the confidence and support by attempting problems in a kind environment that sees the value of their contributions. Kids progress at different rates, and since we aren't training racehorses but instead human beings, a problem-based curriculum can scale to those different rates and students can begin to see their own strengths and contributions. When students see themselves as part of the team pursuing a worthy goal, then the relevance of the mathematics becomes the beacon that they choose to follow.

Further, it's not just about students seeing their own strengths – problem solving and competition math train students to *notice*. Instead of working through prescribed steps to solve exercises, problems require exploration – you need to be in a problem to solve it. That means being able to notice patterns, reframe information, and ask the right questions. Then, the next part of the process is articulating your ideas clearly to your peers and instructor. This process is so critical for every field, especially STEM (science, technology, engineering, mathematics) fields such as data analysis, statistics, finance, and more.

In Section II, "Teaching Problem Solving," we will go into greater depth about the various math competitions and resources available.

Mathematical Relevance: The Math Circle Example

Mathematical relevance is responsible for much of the success of math circles across the globe. In Eastern Europe, math circles have been using this problem-based approach for teaching mathematics for decades. Math circles began popping up in the United States in the 1990s, and there are now *hundreds* of math circles across the country. *The allure of the circle is in reaching out to those who are curious about mathematics, but not necessarily proficient.*

Math circles are math programs for middle and high school students offered on a periodic basis (sometimes weekly, bimonthly, or monthly) and appeal to students looking for mathematics enrichment and topics beyond what the schools offer. Math circles seek to light a passion for mathematics and create lifelong thinkers. Many math circles are open to parents along with the students and draw in mathematicians who love their topic and wish to share their knowledge and interests with all participants. Students are exposed to discrete mathematics, cross-discipline subjects, proof-writing skills, and much more. Lectures can cover number theory, combinatorics, geometry, or algebra, which are typical math competition topics, as well as math-related areas such as art, music, games, economics, history, or physics, plus other topics that can be explored and connected to mathematics.

> *This is a unique feature of the top-tier math circles, not found in middle or high schools, where students are taught to meet state standards on questions that take less than a minute to answer. In contrast, monthly contest problems may take best students hours or days of concentrated thought. Only a few participants are capable of solving all the problems; yet, through the attempt everyone learns about the real world of mathematical research.*
>
> *Zvezdelina Stankova*[1]

Most math circle topics are *accessible* and *scalable* to reach a variety of skill levels so that students not only gain confidence at the beginning of a lecture but also see where the study of mathematics can take them in the long run.

CHAPTER 2: MAXIMIZE LEARNING: RELEVANCE, AUTHENTICITY, AND USEFULNESS 17

 In November 2017, at the Metroplex Math Circle, www.metroplexmathcircle.org, at the University Texas at Dallas, there was a talk titled *The Shape of Space*, by Frank Sottile, a mathematics professor from Texas A&M, who happened to be visiting Dallas that weekend and offered to give a lecture.

The description of the lecture was the following:[2]

> In mathematics and science, we often need to think about high (three or more) dimensional objects, called spaces, which are hard or impossible to visualize. Besides the question of what such objects are or could be, is the problem of how we can make sense of such spaces.
>
> The goal of this activity is to give you an idea of how mathematicians manage to make sense of higher-dimensional spaces and relate this to the recent proof of the Poincaré conjecture that won the Millennium Prize of the Clay Mathematics Institute. We will do this by exploring the simplest spaces, and through our explorations, we will begin to see how we may tell different spaces apart.

This topic was loved by all the participants, for the following reasons:

- Relevance – he connected the topic with the proof that won the Millennium Prize.
- Manipulatives – he used bagels and belts to help with the visualization of the shapes.
- Scalable – the talk resonated with middle and high school age students.
- Fun – he kept the talk interesting and interactive for everyone.

First and foremost, a math circle is a social event where students can engage in exciting and relevant mathematical topics taught by passionate mathematicians who are eager to share their knowledge in an interactive and collaborative environment. Successful math circles happen when the mathematician acts as a facilitator in a student-centric environment where kids are given the time to think deeply about problems and share their knowledge.

Teachers also need to have their passion for mathematics ignited, and to help in this endeavor, there is the Math Teachers' Circle (MTC, www.mathteacherscircle.org) where educators can connect and share ideas to help with professional development and math pedagogy. When teachers tap into a professional learning community (PLC) such as the MTC, they will be able to trade ideas and *been there, done that* experiences to streamline and improve their classroom management as well as their lesson plans.

Why not utilize a math circle approach and break free from rubrics, if only for a certain amount of time, and help ignite a student's curiosity for mathematics by offering topics outside the standard curriculum. This is an opportunity to make connections, explore, and find the story of mathematics.

There are so many incredible mathematicians in every community who are willing to donate their time and energy to share their love of the topic with young students. Speakers can be found at local universities, education businesses, and even math PhD students who are happy to pay it forward with young mathematicians. With the internet, distance is no longer a factor, and a quick video call can be set up to bring this valuable resource into the classroom.

Go down deep enough into anything and you will find mathematics.

Dean Schlicter

 Take time to explore other areas of mathematics such as logic, puzzles, game theory, Fibonacci numbers, magic squares, set theory, math and the Rubik's cube, Conway's game of life, math card tricks, knot theory, infinity, KenKen, geometric constructions, and on and on and on!

There are so many rabbit holes to go down, and that is the joy of the math circle – taking the time to go down the hole with a guide who is just as curious and excited by the discovery as their students!

Curriculum Relevance

Learning the story behind discoveries also brings mathematics to life and makes it relevant. Explaining to students that math breakthroughs came about so that real problems and big questions could be answered. Whether it was dividing property fairly, understanding the calendar, trying to bring order to the universe, or wanting to count the flowers in a garden, math has a story and history that can make learning exciting. Did you know the Egyptians multiplied numbers by doubling? Or that zero started as a placeholder and when it changed to be a number, it elicited great fear – how can you give substance to nothing? The advent of zero allowed algebra to be possible in the ninth century and physics in the seventeenth century. This story is fascinating to students and teachers alike and adds depth to the process.

Studying the biographies of mathematicians can make math relatable as well. What sparked their interest? What was their childhood like? What led to their big discovery?

Everyone has heard that Albert Einstein wasn't a very good student – it makes him vulnerable and achievement seem more attainable. Gauss, Erdös, Ramanujan, Archimedes, Des Cartes, Fibonacci, Pascal, Euler, Hypatia, and on and on – they all have a story that can resonate with students, inspire them to stretch their problem-solving muscles, and provide foundational concepts.

CHAPTER 2: MAXIMIZE LEARNING: RELEVANCE, AUTHENTICITY, AND USEFULNESS

These connections, explorations, and stories bring depth and wonder to the study of mathematics and help engage all students in the class by capturing their curiosity, as opposed to setting up situations where they, instead, feel ranked, such as by timed tests, worksheets, and rubrics.

This type of pedagogy is human-centric and not focused on rote learning. Having some days or time in class set aside for creative design approaches to curriculum is sure to keep students engaged. That doesn't mean it is a free-for-all! Fundamentals need to be covered in a given school year, but there are ways to maintain foundational learning to get them through grade-level requirements (short view) while maintaining their love and interest in the topic beyond what is expected in the school curriculum.

Suggestion Box: Many times, if there is an interesting puzzle or logic problem on the board that kids can work on when they arrive, it primes their brains to be ready for whatever may follow.

Making the problems pertinent to what is happening in their lives (e.g., holidays, social events, etc.) creates more connection and interest. Even rotating through your students' names to use in the problems is a fun way for them to feel connected – it doesn't always have to be Alice, Bob, and Carl.

What Is Your Personal Approach to Problem Solving?

Dr. Mirroslav Yotov: When I teach problem solving, I need to have something interesting to share with the audience. This may be an interesting and unexpected idea, or a problem with interesting formulation, or a series of problems that reveal an interesting aspect of math objects. It is very important to me to give the right motivation for considering the suggested problems. That's why I always strive to give the context for how the problem arises and explain the significance of that problem. Years ago, I was enthusiastic in doing problems where the solution was clever, unexpected, and beautiful. I didn't care too much about the context and the motivation. Now I think that math is not only problem solving, but also an art of asking math problems. And for this, motivation and perspective/context are very important. That is in a sense more important than having the bright idea for answering the question: the right questions prompt interesting ideas! For the motivating approach, I usually start with introductory and explanatory (not necessarily trivial or easy) problems. I gradually steer the students toward the interesting questions to be asked. This way, they themselves become active parts in the process of discovery. This is a longer process, but a better one when novices are taught. I check how the students fare in this process of problem solving by the discussions we have along the way, and by the success they have in doing other, related, problems.

The following Think Tank is a counting problem (i.e., combinatorics, a topic in discrete mathematics) that uses the *principle of inclusion–exclusion,* and if the kids use logic, they can figure it out. It's important to let them discover the solution in their own way. Let them be creative. If they still need the manipulatives (pictures of the animals), they can cut them out of paper and keep track. The important thing to stress is the journey of discovery, looking at what works and what doesn't, and having fun. The next day, they can share their solutions and approaches with the class.

A fellow teacher took a survey of his students to determine the kinds of pets they have at home and discovered that each student has at least one pet. He tells you that 20 students have a cat, 15 students have a dog, and 7 students have a bird. If 2 students have a bird and a cat, 4 students have a bird and a dog, 6 students have a cat and a dog, and 1 student has all three pets, how many students are in his class?

Here is the solution using a Venn diagram. It's easiest to create the diagram working backwards with the student who has all three pets. Now, it is clear that he has 31 students in his class.

How does this work mathematically? 20 students have cats + 15 students who have dogs + 7 students who have birds = 42 pet owners. But wait, if you just add the pets, you double count the students who own two or more, so they need to be subtracted out: 42 – 2 – 4 – 6 = 30. The answer is closer to being correct, but something is still wrong. Why? It's because of the student who owns a bird, a cat, and a dog who was subtracted three times instead of twice, so he needs to be added back in to equal 31. The following Venn diagram shows how many times each group gets counted when we add the total members of all three sets:.

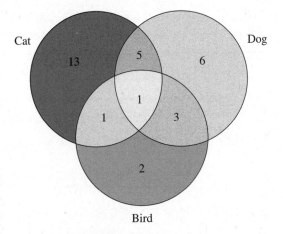

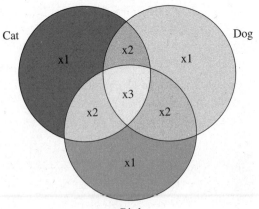

This problem is a great exercise in thinking while also opening the door to more difficult forms of combinatorics (counting) problems. Therefore, it is scalable, depending on the needs of the students and fun to work on collaboratively.

And as in life, balance is important, and that means ensuring that kids are offered physical mathematical enrichment (e.g., movement, manipulatives, and group work), as well as mental mathematical enrichment. The *Sage on the Stage* approach is lecturing from a central point to your audience, but with design thinking, you can mix it up!

Keeping a sense of discovery and newness in mathematics is critical, and problem solving, with interesting challenges, ensures that the thrill stays alive. Whenever learning becomes stale, rote, check the box, or – *gasp* – timed, then brains turn off and students become trained to just go through the motions.

 Create math stations where students physically move to learn new concepts, work in groups, or are reintroduced to manipulatives that solidify understanding. Math games, problem hunts, and mathematical modeling lessons are all examples of ways to add physicality to the classic design of the math class.

Authenticity: The Cargo Cult Science Trap

Authentic learning means more than just going through the motions; you don't want your students to fall into the cargo cult science trap. The famous physicist Richard Feynman discussed this trap in his 1974 commencement speech at Caltech. He brought up what he called *cargo cult science*. Cargo cults are part of religious practices where tribal cultures who have not been exposed to industry or technology attempt to mimic the motions necessary to receive the cargo, or bounty. During World War II, a tribe in the South Seas witnessed cargo being dropped by airplanes onto islands where temporary air strips had been built. The soldiers would share the cargo with the tribes, who grew accustomed to these new delicacies and treats. When the war ended, and the air strips were dismantled, the tribes still wanted to continue to receive the material wealth of the cargo. They imitated the actions that they believed made the cargo appear out of bamboo and local materials. They built landing strips, airplane models, and headphones, all from bamboo and wood, to mimic the motions of what they observed from the advanced culture. And while the cargo cult might seem to have been doing everything needed to bring bounty from the sky, they weren't truly understanding what was actually happening.

They're doing everything right. The form is perfect. It looks exactly the way it looked before. But it doesn't work. No airplanes land. So I call these things cargo cult science, because they follow all the apparent precepts and forms of scientific investigation, but they're missing something essential, because the planes don't land.
  Richard Feynman

Dr. Feynman was stating that the missing piece was the integrity, truth, and honesty that you are working to find a solution to a meaningful problem instead of following a bias or mimicry or glory. Students, to truly be educated, need to be just as interested in what doesn't work as what does work. Fear of failure cannot be an option because discovery is all about the honest testing of ideas and finding a path, pattern, or process. Dr. Feynman was an advocate for understanding the beauty and depth of problems, the underlying *why*, so that truth can be discovered.

Some students know the steps that need to be taken to find the square root of a number, but very few have understood *why* you would need to do so. Square roots take off the training wheels of arithmetic and lead young mathematicians to more abstract mathematics and algebra. First, they learn that $x^2 = x*x$ is the area of a square of side length x. If discussing a non-negative length x, then $x = \sqrt{x^2}$. Squares are a common geometric shape for students to start with, and multiplying $x*x$ is a first glimpse into exponents and more areas to investigate.

Authenticity in Learning

As an educator, guide, and mentor, your main job is to facilitate authentic (real) learning and teach students to *know* when they are on the right track.

To accomplish this goal, you can follow the same methodology used by doctors to diagnose wellness.

1. They want to understand the problem.
2. They look at the patient's data (test results) and listen to their concerns.
3. They make a diagnosis and treat.
4. They follow up with the patient to see if the problem has been corrected, and if not, they start again.

Therefore, it's not only important to facilitate learning but also to help develop the students' mathematical compass so that they can begin to self-diagnose issues and remain authentic to the process. In the end, there is only so much time in a day, and as an educator you don't want your students lost in the woods of a complex problem. How do you give them the compass to know if they are going too far off trail?

Help Your Students with This Five-Step Self-Assessment Process:

1. *Do you have the foundational knowledge to solve the problem at hand?* There is a large difference between knowing and understanding. When a teacher is the one solving a problem on the board, it can seem so easy to do, but when the students must solve a similar problem independently, they can get stuck. That's because real understanding takes longer and requires the students to truly understand what the underlying concepts are to complete the work.

2. *Do you understand the problem; what information is given, and what is being asked?* More times than not, students will veer off path because they haven't clearly read the problem. They can get excited because they will partially read, think they know what the problem is asking, and will impetuously start solving. They need to carefully read the problem, underline key words, and then start with what they truly know.

3. *Do you have a strategy for solving the problem?* Once step 2 is complete, it's time to devise a strategy. This strategy will provide the signposts necessary to know where you are going and how you arrived where you are. In the heat of the moment, you may not write down every step to solve the problem, but you should have enough signposts to let you know where your brain is taking you.

4. *Does your strategy work?* Test out your strategy and see if you come up with a solution. If the strategy doesn't work, then go back to step 3 and amend your approach.

5. *Is your strategy, solution, and process effective and/or efficient?* It's always important to self-reflect and see what was successful and what needs more work.
    a. Did you use your entire strategy (hypothesis), or just part of it?
    b. Can you formulate a more general question?
    c. Have you given all the information necessary (steps) for someone else to understand your solution?
    d. At this point, collaboration is critical. Students can use their teacher and classmates as feedback loops to see other approaches used for the problem and collaborate to see what is most effective.

Five Common Mistakes Made by Every Problem Solver

1. Not understanding what is being asked/required
2. Not reading carefully (missing information)
3. Not solving slowly when thinking quickly
4. Not employing feedback loops
5. Not checking their computations (simple errors cause a lot of issues)

If given the proper foundational understanding, meaningful problems that can illustrate this understanding, and the tools to know if they are on the right track, students will imbibe the information and it will stay with them much longer. Peers are also critical in this endeavor so that ideas can be discussed and approaches explored. This is why complex problems that can be solved with multiple approaches can make such a huge difference in learning.

Here is a complex problem that is sure to connect with the kids, at least with the kids who have cats!

A kitty named Meow jumps on a computer keyboard that only has 26 keys corresponding to the letters of the English alphabet. Meow's paws land on four different keys. What is the probability that the keys Meow jumped on can spell its name?

Solution 1: We only know that Meow pressed four *different* keys. This tells us that the total number of possible four-letter words that these keys can spell is $26 \cdot 25 \cdot 24 \cdot 23$. Out of this number, only $4 \cdot 3 \cdot 2 \cdot 1$ words consist of letters M, E, O, and W. Therefore, the probability is

$$\frac{4 \cdot 3 \cdot 2 \cdot 1}{26 \cdot 25 \cdot 24 \cdot 23} = \frac{24 \cdot 1}{26 \cdot 25 \cdot 24 \cdot 23} = \frac{1}{26 \cdot 25 \cdot 23} = \frac{1}{14950}$$

Solution 2: Recall that Meow pressed different keys, so it is randomly picking a four-element subset from a 26-element set. Only one subset, {E, M, O, W}, is favorable, so the probability is

$$\frac{1}{\binom{26}{4}} = \frac{1}{\left(\frac{26 \cdot 25 \cdot 24 \cdot 23}{4 \cdot 3 \cdot 2 \cdot 1}\right)} = \frac{4 \cdot 3 \cdot 2 \cdot 1}{26 \cdot 25 \cdot 24 \cdot 23} = \frac{1}{26 \cdot 25 \cdot 23} = \frac{1}{14950}$$

Solution 3: Yet another solution is based on what each of Meow's paws has pressed, in any particular order. For example, starting from the front left paw for a favorable outcome, it had a choice of 4 letters out of 26. The front right then had a choice of 3 letters out of 25, the rear left paw had a choice of 2 out of 24 letters, and the rear right had only one favorable choice out of 23, for the final probability of

$$\frac{4 \cdot 3 \cdot 2 \cdot 1}{26 \cdot 25 \cdot 24 \cdot 23} = \frac{1}{26 \cdot 25 \cdot 23} = \frac{1}{14950}$$

Usefulness

Learning needs to be useful as well as relevant and authentic. Why are there so many math enrichment centers popping up across the country? When students spend an average of six to seven hours per day in school, why would they need additional math enrichment? When you look at the types of enrichment centers, they serve the following purposes:

- Tutoring centers help struggling students get up to speed.
- Advanced mathematics programs teach concepts not offered in today's schools.
- Strategy centers teach problem solving through games, puzzles, and logic.
- Competition math training challenges students and prepares them for discrete mathematics, college admissions, and critical thinking.

Students who cannot afford these extra enrichment opportunities are left behind. This doesn't have to be the case. Schools have dedicated teachers who can incorporate what makes these enrichment centers valuable to parents and students by offering a useful, problem-based program that is scalable, engaging, relevant, and fun.

Notes

1. Zvezdelina Stankova, *A Decade of the Berkeley Math Circle: The American Experience* (MSRI Mathematical Circles Library) (v. 1).
2. Frank Sottile, "The Shape of Space," lecture at the University of Texas, Dallas, November 4, 2017.

CHAPTER 3

Creating a Math Learning Environment

Before a problem-based learning approach can be implemented, the environment has to be conducive to making it work, which means the students, teacher, school, and district must be able to work together in a cohesive way. Teaching the next generation of mathematical thinkers requires introspection, analysis, understanding, and a plan.

As the educator, you need to:

- Know yourself.
- Know your students.
- Know your approach.

Mathematics is, in many ways, the most precious response the human spirit has made to the call of the infinite.
    Cassius J. Keyser, The Human Worth of Rigorous Thinking: Essays and Addresses

Know Yourself: Ego and Grace

As an educator, beyond your own goals for a class and personal growth expectations, there are a lot of external pressures and expectations for achievement and progress:

- Individual students' progress
- Parent expectations
- School expectations
- District and state expectations

With that in mind, the educator should consider two important personal mindset aspects when coaching mathletes: (i) *ego* and (ii) *grace*. Ego tends to be more of a fixed mindset approach (authoritarian) when you think you should be able to best the students because of your experience and education and they, in turn, should fall in line. Grace in this book is defined as self-forgiveness when the lesson doesn't go as planned and accepting mistakes as an opportunity to pivot and try a different approach.

EGO

When engaging in a team atmosphere approach to math education, it's paramount to check your ego at the door. Young problem solvers can be fast, creative, and take a problem down a path you haven't yet conceived.

As Dr. Andreescu has noted in his vast coaching experience:

> *It's like you are in a tennis match with the student; they may be faster and more agile, but you know where the ball is going.*

While it will be a challenge to keep up, *your* math education is stronger, and you can learn new things in the process. It's also good for kids to see you struggle, evaluate a new approach, and enjoy the journey with them. It's not about always having the answer; it's about being a part of a team that is working together to understand the heart of a problem – it's a quest for truth.

This following problem is an enticing visual that promotes class discussion and inquiry:

What could be the next term in the following series of objects?

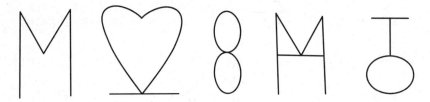

Solution: If we draw a line as in this image:

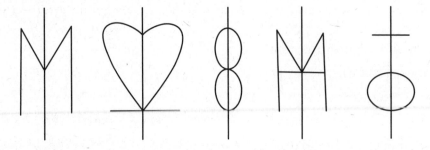

And remove the left half of each object, we obtain

1 2 3 4 5

Hence, the next symbol is[1]

When there's a playful atmosphere and people are solving problems together, they get so much farther in learning than if they are following a regimented and inflexible authoritarian approach. It's important to keep in mind that if you're not enjoying doing the work or the process, then neither will the students. Try new things and give yourself *grace* when you need to pivot and change your plan.

GRACE

Ego is the hubris or arrogance that you *know all* and it decreases your flexibility, whereas on the flip side grace is the humility to understand that you are fallible. As an educator, there are always ideals to be reached when designing lessons, but they aren't called *ideals* for nothing – and many times we fall short of them. This is where grace comes into play and giving yourself permission to try new things, fail, get up, iterate, and try again in a different way. There are seasoned educators who seem to have it all together, but remember, what may appear to you as seamless was probably hard earned and it doesn't mean it will continue to work over time.

This is where joining a community of teachers helps so that you have a place where you can share, learn, and innovate. When you engage in the trade of ideas, everyone improves. This collaboration can help educators just as much as it helps students in the class.

Flexibility, creativity, patience, and persistence along this educational journey are just as important for the teacher as they are for the student, so be sure to celebrate your successes, learn from the failures, and keep striving for the ideals while giving yourself the grace and forgiveness to accept when a lesson doesn't go as planned.

Know Your Students

As a teacher, besides utilizing and/or designing engaging curriculum, it is also important to design your class in such a way that students work in groups that bring out their strengths and not their weaknesses. We all have different learning styles. Some are complementary and some are not. Students must learn how to work in both types of situations, and much of that is knowing themselves well and knowing what value they add in different types of group settings.

There is an old joke:

> When I die, I want the people I did group projects with to lower me into my grave, so they can let me down one last time.

So how do you help students to know *how they think* so that they know what value they add with collaborative problem solving? There are a lot of tools out there for doing this, Myers-Briggs or Howard Gardner's Frames of Mind to name a couple, but in the end, a simple approach can also be useful, and that's taking the time for students to think about *who they are* and *what they like to do*. You can help your students *find their Venn*.

FINDING THEIR VENN

Finding their Venn is a lesson in introspection. Our inner cores stay remarkably the same. Before life became about homework, video games, social media, and/or meeting expectations, how would students spend their free time when they were around seven or eight years old? Building things? Making up stories? Studying dinosaurs? Playing outside? Reading books? Working on puzzles? Talking about the world? Sports? Music? Playing games and if so, what kind? Exploring ideas? Was there a topic they liked more than others? The trick with this exercise is to look at the activities a child was engaged in, and then extrapolate a core value. This is an incredibly difficult exercise, so students might need to consult with their parents to help determine what drove them when they were younger, but it's worth the time so that your class can be designed with the students who you are teaching in mind.

Here are a few types of kids and their interests to get you thinking:

1. The child who is making up worlds, whether they write stories or design games or play with languages, is a creative kid. In music, they may enjoy composition more than perfection with playing a piece. In mathematics, they would benefit from seeing visual proofs or deriving formulas or creating their own problems for the class to solve.
2. Everyone knows a *"dinosaur"* kid, one who loves to study and immerse themselves with all the facts about dinosaurs, or the *baseball* kid who memorizes all the stats of players, and so on. These kids are immersive learners and want to *collect* mathematical techniques and tricks. They prize efficiency in problem solving and tend to work through large amounts of problems at a time.
3. A student who was the *why* kid, "Why is the sky blue?," "Why do people get sad?," "How do bumblebees fly?" may be the type of child who wants to know the story of mathematical concepts such as those attributed to Pythagoras.

CHAPTER 3: CREATING A MATH LEARNING ENVIRONMENT

Finding your Venn is all about determining what puts you into a *state of flow*. Flow is when you are process versus results oriented. So how do you find those core traits for your students? First, you need to have them stop and reflect about what puts them into flow.

Signs of Being in Flow

- Putting off biological needs, e.g., eating
- Losing track of time
- Energized laser focus that blocks all outside distractions
- Non-stop activity
- Effortless work

***Finding Their Venn* Introspection Questions**

Here are some questions you can ask the student to find out what fuels their passions or flow. They need to think in terms of their younger selves (seven or eight years old).

- What choices did the student make as a child for their free time?
- What things would the parent have to ask their child to stop doing so as to move on with their day?
- How or when did the student truly light up? Through certain activities or discussing ideas or both?
- What activity(s) would make the student lose all track of time?
- At what times did the student show real confidence and love for what they were doing?
- When did the student seem the most intense?
- If the student was stressed when they were younger, what would be suggested as something they could do to take their minds off it?
- What times would the child's mind be working so fast that they couldn't communicate quickly enough?

Here are a few example diagrams from previous students:

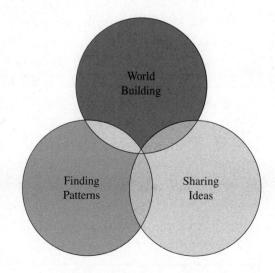

Student A. As a child (and it still holds true), this student loved to create imaginary worlds and lived a lot in his head, which is easy to understand, because it was a fascinating place with so much going on. If he read a book or story, he would incorporate the aspects he liked the most into whatever story or game he was building or even go so far as to write *fan fiction* and continue the story even after the original ended. He would seek out patterns and connections, so was particularly drawn to math and would find other friends to which he could explain new and interesting concepts. Writing problems that others would enjoy solving brought joy as well, and this transferred to music, where he would compose songs and loved writing lyrics. Now, as an adult, his passions are pure mathematics, theoretical computer science, and creating video games.

Student B. This was the kid who would ask why people were crying, what happens when you die, what is the last connection you would want to make with someone. He was a truth seeker and had a high *justice meter* (people needed to behave rationally and with honor). That said, he was also mischievous (and this is meant in a positive way) and would seek out the edges of systems and try to unravel them. He would process information through movement, constantly running in circles around those he would want to explain his ideas to. He has high energy and a large heart. Today, as an adult, he enjoys CrossFit, economics, philosophy, and applied mathematics.

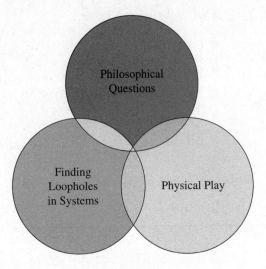

Student C. She would completely lose herself in a topic and consume whatever was possible to learn about the subject. This meant bridging that learning into other areas and seeking out the connections that led to understanding as well as contributed to the greater fabric of knowledge. She would connect with people over ideas (as opposed to emotions) and felt that those relationships were the most worthwhile. Nothing would make her light up more than a new perspective when thinking about a topic, and she could immerse herself in the tiniest of details. As an adult, she enjoys research, education, and cross-discipline pursuits.

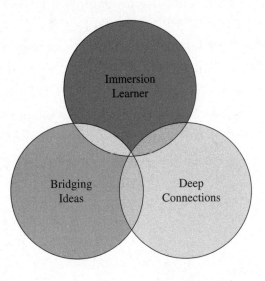

Pythagorus's story is full of interest and intrigue, For example, he was convinced the whole universe was based on numbers, and as such, the planets and stars must move based on mathematical equations. He felt these equations corresponded to musical notes, which produced a kind of symphony, the "Musical Universalis" or "Music of the Spheres." The biographies of famous mathematicians will bring theorems and concepts to life for a student of this type. (Check out the full units on Triangular Numbers and the Pythagorean Theorem Revisited to learn more!)

However, just like Pythagorus felt that the number 3 was an ideal number because it has a beginning, middle, and end, therefore representing harmony, it's best if the students reflect and come up with at least three core values to ascribe to themselves so that their true complexity of thought and uniqueness can shine through.

You can also observe for yourself how a student handles a new challenge in class. When faced with this challenging problem:

- What is their approach?
- Do they employ a specific technique to solve it?
- Do they stand up and move around so that they can think?
- Do they look for patterns or start writing down ideas?
- Do they think about similar problems?
- Do they ask for more clarification?
- Do they ask for the story behind the problem or more facts about it?

When working with young mathletes, they tend to shift along a spectrum of efficiency and creativity, some skewing more closely to one end of the spectrum than the other.

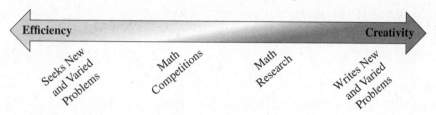

Those who prize efficiency love to solve problems quickly, so they can work on more problems! They follow a methodical and linear progression to problem solving and tend to have incredible memories. Those who prize creativity want the full story and tend to think in three dimensions. They want to pin that knowledge in the fabric of their creative brain space by connecting it to other ideas, finding the story behind it, or deconstructing/reconstructing the problem. They aren't as fast to memorize, but once they have the idea firmly placed in their *fabric*, they remember it longer even though it may take longer to learn. Again, this is a spectrum, not an either/or, so understanding the *how* behind how a child learns allows for more patience and understanding when moving forward.

This approach is more closely related to what students will face in the real world when they go into businesses, academia, or other organizations. Are they a Hipster (designer), Hacker (how it gets done), or Hustler (how it is sold)?[2] We all have to figure out how we learn best, how we solve problems, and where we add value. Why not begin this process in mathematics courses? Students will see that what they are learning carries them past their time in school and will benefit them throughout life.

VENI, VIDI, VICI (I CAME, I SAW, I CONQUERED)

To be an exemplary math coach, it takes meeting students where they are and knowing how they learn (*Veni*), training them in the mathematics and problem solving necessary to thrive both while in school and out (*Vidi*), and providing them with the confidence to add value to the world (*Vici*). This, of course, can be the goal of any subject; however, mathematics is an exceedingly efficient way to learn the inductive and deductive reasoning necessary to make smart decisions and solve problems.

Inductive reasoning problems start with observations of patterns or trends and then generalize that data to arrive at a conclusion or conjecture:

> Last night there was a party, and the host's doorbell rang 20 times. The first time the doorbell rang, only one guest arrived. Each time the doorbell rang after that, two more guests arrived than had arrived on the previous ring. How many guests arrived at the party?
>
> **Solution:** The situation can be represented as a sum of consecutive odd integers starting from 1, which we know to be the square of the number of terms:[3]
>
> $1 + 3 + 5 + \ldots + 39 = 20^2 = 400$

Deductive reasoning problems are when you deduce conclusions from given facts you know are true utilizing logic, as in the following:

> The diagram shows some squares whose sides intersect other squares at the midpoints of their sides. The shaded region has a total area 7. Find the area of the largest square.
>
> **Solution:** The answer is 56. The shaded region has area equal to that of the center square. Each square has an area that is half the area of the square that surrounds it. It follows that the largest square has an area that is 8 times as large as the shaded region. Therefore, the requested area is $8 \times 7 = 56$.[4]

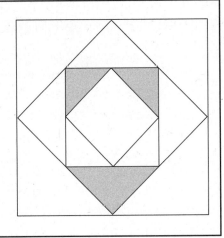

Reason, logic, facts, conjectures, theories – they all roll together to create tomorrow's thinkers! Students who know how to evaluate research studies, statistics, medical reports, business plans, and so on can look at the data, ask questions, and find truth.

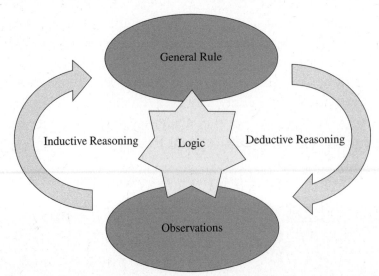

The goal of coaching is to allow each student to have the ability to say *Veni, Vidi, Vici* – in other words, I know how I learn, I know how to solve today's problems, and I have the confidence to conquer other problems in my path.

Know Your Approach

LEARN FROM THE SYSTEMS THAT ARE SUCCESSFUL TODAY

A number of widely successful and talked-about systems, mindsets, processes, philosophies, approaches (it's difficult to come up with one name or classification) are making a large impact in education, business, software development, and so on. Here are a few examples:

Growth mindset. "In a growth mindset, people believe that their most basic abilities can be developed through dedication and hard work—brains and talent are just the starting point. This view creates a love of learning and a resilience that is essential for great accomplishment."[5]

Agile development. Agile software development refers to a group of software development methodologies based on iterative development, where requirements and solutions evolve through collaboration between self-organizing cross-functional teams.

Design thinking. As defined by IDEO founder David Kelley, design thinking is "a human-centered approach to innovation that draws from the designer's toolkit to integrate the needs of people, the possibilities of technology, and the requirements for business success."

Why are these systems so popular now? What has changed? The common thread is that they are all human-centric approaches. In the past, when the United States modeled its education systems or workplace systems, the limiting factor to productivity was the power of the machine or factory, but human time was relatively abundant. Think of the computer scientist of years gone by, work, work, working on their stack of punch cards that they would then feed into a machine to process overnight. At that point, human time was squandered so as to maximize the efficiency of the machine or factory. However, when computing became abundant and the constraint became humans, this meant that talent and agile responses were more important to productivity, and, subsequently, processes needed to be maximized around human resources and *not* the machine.

This shift also applied to the end-user experience. For example, machines were historically designed around engineering principles (e.g., faster cars or multipurpose machines). Remember the combination VCR plus TV product? It was a huge flop because it didn't consider how the consumer actually used the products. The revolution of design thinking was that *better is in the eye of the beholder* – namely, the end user determines the value of the design. Is it a better design for *me* the individual?

Math education, or all education for that matter, needs to be *human centric* to prepare students for a world where critical thinking, talent, problem solving, and collaborative environments reign supreme.

> *Tell me and I forget. Teach me and I remember. Involve me and I learn.*
> *Chinese proverb*

This human-centric or customized world is the only one in which today's students have had exposure. They didn't grow up with broadcast television where you need to sit in front of your TV at the prescribed time to watch your favorite show. They've grown up in a world of complete customization for the individual. They can curate their own entertainment, decide what fashions look best for their tastes, and build, customize their electronics, and choose their identities based on what feels natural to them.

The only area where they are still exposed to a broadcast approach is education, and students' talent and abilities are subjugated to the machine of GPA, standardized testing, and rubrics that fulfill the goal of accommodating an idealized, nonexistent, *average student*. Instead, math education should be designed for the individual and specifically to those on the edges.

DESIGN TO THE INDIVIDUAL

When you design a mathematics program for the average student, in actuality, you've designed the program for zero students. However, when you design to the ends of the bell curve in the class, you create an active learning environment that allows all students to excel. This concept of *average* in education has been brought to light by L. Todd Rose, director of the Mind, Brain, and Education program at the Harvard Graduate School of Education, where he leads the Laboratory for the Science of Individuality and author of *The End of Average: How We Succeed in a World That Values Sameness*. As he points out, there is no such thing as an average student just like there is no such thing as an average genome, average memory, or average cancer.[6] And yet, our entire education system is built around the average student (e.g., these are the things the average 8th grader should know). Another term used in education for *average* is *standard*. Standardized tests, standard time to learn concepts, standard material.

This *top-down* approach to education does not benefit today's students, or in Mr. Rose's vernacular, an education for the average student is designed for no one. Problem solving allows for a *bottoms-up* approach where the goal is for each student to be better than they were the day before. Students are given the time and tools they need to solve the problem in their own way, they are given the support to follow through on their ideas, and they can be given problems outside their grade level to see the wonders that lie ahead. Problem solving prepares students for the world they will face, which is a continuously changing *nonstandard* landscape that requires dreamers, innovators, entrepreneurs, and in all areas, problem solvers.

This means the class needs to be *flipped* (not a *sage on the stage* model), and a collaborative and synergistic atmosphere should be created. For example, the teacher will offer a brief discussion of a topic and then the students put their brains together to come up with solutions. The teacher is a facilitator and guide during this process, circulating through the room to offer hints and filling in knowledge gaps.

FLIPPED CLASSROOM

When class time is limited, teachers don't want to lose precious collaboration moments with lecturing. While it is a lot of upfront work, creating flipped classrooms where students work on problems during class time and the assigned homework is the lecture can maximize the effect of problem-based learning and collaboration. Further, when a lecture is recorded, students can slow it down, watch it multiple times, and/or skim over areas they know well. Here are some pros and cons of flipped classrooms:

Pros

- **Student centered.** When students drive their own learning by working to understand the lecture, retention of material and confidence can be increased. This method also increases teachers' one-on-one learning with students who may need extra help.
- **Collaboration.** Class time can be used for student and teacher collaboration. When students present problems on the board, the entire class along with the teacher can support their progress and help them with deeper understanding of the material. Working in groups also provides the accountability to participate, helping to ensure that students actually watch the lecture, so they can add value to their group.
- **Parents can be a part of the process.** Since the lectures are recorded, if a student has a parent or friend who can help at home, they can watch the lecture together. It gives parents a *map of engagement* for working with their child and their teacher, meaning that parents can help the student develop what questions they should ask the next day to increase mathematical fluency.

Cons

- **Upfront work.** Initially, creating the lectures can be a lot of work for the teacher, who may or may not teach the same classes year after year. These lectures can be in note form or as recorded videos, where the teacher needs to think ahead and create a complete curriculum to satisfy all learning requirements. Recordings are the most common way to create the lectures; however, not all students may have the same access to the technology needed to view them.
- **Math phobia.** Can create a "fight, flight, freeze" phenomenon where students have a difficult time understanding the lectures at home. It may take a while to get over these fears, making a flipped classroom approach take longer to show results.
- **Pacing requirements.** Set by the district may make it difficult for teachers to have the necessary time to effectively start a flipped classroom approach. The lecture model is a faster way to disseminate information to a large group of students, even if it isn't the most effective for long-term retention or increasing test scores.

The curriculum included in Section III of this book is intended to mitigate these cons by providing the lecture notes and scaffolded problems.

JOURNEY OF DISCOVERY AND THE IMPORTANCE OF RISK

With a human-centric, individualized approach to education, it's important to ensure that the student is the *hero* of their own story. Joseph Campbell created the template for the Hero's Journey or Monomyth. He was a literature professor whose research was in comparative mythology and comparative religion. In this template, the hero has a *call to adventure* where she must decide whether or not to *cross the threshold* to the unknown. On her adventure, she receives aid from helpers and mentors while she works through trials and challenges. Eventually, she has a *revelation* and is transformed by the process, returning once again back over the threshold and sharing the *elixir* of what she has learned.

As the hero of their problem-solving journeys, students will go through these transformations multiple times, and their mathematical knowledge will deepen and transform. As they grow in knowledge, they can become the mentors and helpers for others on the path.

The construct of the Hero's Journey (a.k.a. Monomyth) applies well to those studying problem-solving and mathematics. The threshold to adventure should be an interesting problem or challenging concept that is worth pursuing. In the Monomyth, the threshold can be as simple as going into a dark and mysterious cave where you have adventures and come out the other side.

Even this simple experience can be a wonderful problem-solving opportunity.

This logic problem is exactly what an adventure should be: interesting, engaging, and having depth and layers. This is a common type of problem and can help build problem-solving skills for each student. However, as stated above, the hero is not alone, and she has helpers and mentors who can guide her along her path.

Let's say that the threshold is crossed, and our hero must traverse many twists and turns when, all of a sudden, the cave splits into two paths. One path leads to certain death and the other to freedom. Each path is protected by a guard. One guard only tells the truth and the other guard only tells lies, but you don't know which is which. You can ask one question; what would it be?

Solution
What cave would the other guard tell me to take?
Explanation: Whatever cave the guard tells you to take, you would take the opposite. The guard that always tells the truth would lie because that is what the other guard would do. The guard that always lies would, well, lie and not tell the truth. Therefore, you would take the opposite cave.

This is how humans flourish, by engaging with their community, sharing experiences, and working on hard problems. Adventures aren't always appealing at the outset. That's why in the Hero's Journey, the hero will initially refuse the call to cross the threshold; they feel like they aren't the right person nor have the right skills. It's a big ole case of Imposter Syndrome – not feeling accomplished enough to tackle the problem. Students must feel worthy at the outset, and this happens when they are in a supportive environment with positive guidance. Then, they will cross the threshold, have the confidence to solve problems, and be more willing to take intellectual risks.

A man's errors are his portals of discovery.

James Joyce

Intellectual risks are what stretch us to be better – leaving the comfort zone of our front door and embarking on new quests. But we all bring different skills to the table, and students are jagged learners. Some may be gifted in math but socially awkward. They may be socially connected but have trouble with focus. They may have great focus but are slow with reading comprehension. And all the variations in between and then some. That is why taking intellectual risks in a human-centric classroom is so important. Failure is a critical learning tool, and if students can learn how to *fail, then they can learn how to fail faster, meaning they will gain insight on shortening the fail time as well as learn how to pivot quickly enough so as to avoid hard falls.*

Because stakes have been raised as far as going to college, college admissions, and false metrics (e.g., grades, GPA, standardized test scores), students don't want to take intellectual risks. How many times have you heard, "Will this be on the test?" or all the variations thereof, such as "Can you tell me what will be on the test?" or "Do I really need this?" The shift has focused from intellectual query and risk to playing it safe.

That's where a problem-solving curriculum can help. When grading a problem-based curriculum, you aren't focusing on the end product, you are focusing on the thought process (the journey). When your main concern is each student improving every day, you can evaluate how that progress looks for that particular kid. Further, flipping the classroom will allow the collaboration necessary so that students can see their own mistakes and have feedback loops in place with you as the mentor and their peers as their helpers.

Here is a problem where you can grade the student easily based on how they approach this problem coupled with their steps to solve it.[7]

From a class of 30 children, we need to choose 10 who will represent the class at a math competition. In how many ways can we do that if:

1. There are no limitations?
2. Student A said that he will participate only if his best friend B participates, too?

Solution:

1. The number of possible teams is

$$\binom{30}{10} = 30045015$$

2. One way to solve this problem is to count all possible teams when A participates (and therefore B also participates) and separately if A does not participate:

$$\binom{28}{8} + \binom{29}{10} = 23138115$$

Another way to solve this is to count the number of teams in which A participates but B does not and to subtract that number from the total number of teams:

$$\binom{30}{10} - \binom{28}{9} = 23138115$$

How Have Your Teaching Methods Evolved Over Time, and Why?

Dr. Branislav Kisačanin: Over the years, I became increasingly confident that inspiring students is more important than lecturing, as in letting students present their solutions in front of their peers, instead of me presenting all the problems.

Furthermore, having students present improves their confidence and presentation skills, which are so critical in the modern world. When I started working with mathematically gifted kids, I would spend approximately 20% of the time on inspirational stories, 0% of the time letting students present, and 80% of the time on lecturing. Fifteen years later, these percentages have shifted to closer to 40% inspirational stories, 30% letting students present, and 30% lecturing, and I'm continuing to evolve in this overall direction.

Heroes are the protagonists in the story, which means they need the opportunity to be front and center. An important aspect of a problem-based curriculum is having students present their work in front of the class. For many students, this is a very scary adventure threshold to cross, and as their guide, you need to know how best to inspire your students to take this step.

MISSION STATEMENT AND CORE VALUES

Why would a teacher want a mission statement and set of core values created for their classroom? For the same reason businesses need a mission statement and core values:

- To let their stakeholders (the students) understand what the class is trying to accomplish
- To help the student choose their role in achieving these goals
- To create an environment conducive to learning and growth toward the mission

Mission Statement

Mission statements set top-level goals for the class, provide a cohesive team environment, and allow students to assess their contribution to the group. And as always, we want students to be ready for the world in which they will be entering; therefore, early exposure to tried and true business practices will give them the purpose and connection to something larger than themselves – a critical component to tapping internal motivation.

The AwesomeMath mission statement allows us to tell our customers, parents, and educators what types of students are a fit and what products/services we offer, and it sets the tone for the environment and community we like to create:

> *AwesomeMath is devoted to providing enriching experiences in mathematics for intellectually curious learners. Through summer camps, publications, curriculum, and competitions, AwesomeMath fosters a community that values critical thinking, creativity, passionate problem solving, and lifetime mathematical learning.*

One of the main reasons a business will have a mission statement is for its stakeholders to understand what the company is all about. A classroom mission statement strives to accomplish the same goal and allows your students (the stakeholders) to understand the purpose of the class as well as where your heart is when teaching it. Further, the mission provides a reflection point throughout the year where the class can provide feedback as to whether they are on task to reach their mission or not. This means, you must also be open to the mission evolving and changing as the needs of the stakeholders evolve and change. *As with all things in this book, it is about the process and not top-down directives, so buy-in from the stakeholders where they can contribute to the mission will have a more positive outcome of reaching the goal than an authoritative missive meant to result in compliance.*

What are the components of a good mission statement? It should be clear, concise, and connect with all stakeholders.

> **Clear.** *What a*re the objectives of the class? What value does the class add to each student? You *don't* want to list specific items, e.g., "This algebra class seeks to provide understanding in the areas of patterns, linear equations and inequalities, statistics, exponential equations, and quadratic functions," but instead describe what is the common thread that unites all these areas, e.g., critical thinking, logic, creativity, etc.
>
> **Concise.** *What m*ission can these students easily unite behind? Is there a focused and discernable direction apparent in the statement? It's not about being excellent at specific concepts, but instead, showing the students to what they are connected and how they will evolve.
>
> **Connect.** *Student*s want to be a part of something bigger than themselves, be inspired, and have a lifelong growth mindset. The mission needs to capture these goals so the stakeholders unite and connect to achieve them, heart and mind, and feel they can make a significant contribution.

Here are sample mission statements that could be created by your students depending on their maturity levels; the first is more complex and the second more simple:

- This mathematics class rewards the process over the outcomes so that each student can grow mathematically, be innovative and curious, and collaborate together to solve meaningful problems that will help them succeed in school and prepare them for the world ahead.
- Challenging mathematics in an environment that is kind, collaborative, and respects each student.

Again, it works best if everyone works together on the mission, with guidance from the instructor, so they are invested in the statement *and* the outcome. So while you may like components of the mission listed, your own class, school, district, and personal style will shape the mission to be most relevant to you and your students.

Young students have so many unknowns ahead of them in life. As adults, many of our big-life questions have been answered, the most important of which is a clear sense of purpose or self. That's not to say that when you reach adulthood, *KAPOW!* you now have all the answers to life. More that you are farther ahead on the journey than your students and therefore have the opportunity to act as a mentor/guide for the kids in your class. The mission statement can help set them on a clear path to achieve the requisite goals.

Guide your students as they embark on their *world of unknowns* by providing a framework for learning and life in your class. When you're in a mentorship role, it's never one-dimensional and it's never just one topic. You're serving as the navigator that provides direction to their lives, and they're going to model themselves after the attributes they admire in you and other guides along their path.

Providing a mission statement that is *clear*, *concise*, and *connects* will help add clarity to the unknown of where their mathematics course will take them, and it will give peace of mind regarding expectations, as well as provide the instructor a way of assessing the success of the class. While the mission statement sets the course, a list of *core values* sets the behavioral expectations for the journey.

Core Values

Having core values allows the instructor to create an environment conducive to learning and provide a guide for students so they can self-check their behavior and attitude in a way that is not personal. This has been an invaluable tool at the AwesomeMath Summer Program where the core values are:

- Kindness
- Community
- Independence
- Fun

Whenever a student strays from the core values of the class, they can be asked, for example, "Are you demonstrating kindness in this situation?" This way, they can self-assess their behavior and choose to correct it, especially if they are a part of designing the values from the start. The list of values should be short and sweet, and students can work with their instructor to find the words that they feel will bring out their best.

Here are some value words to get you started:

- Kindness
- Collaboration
- Creativity
- Honesty
- Innovation
- Excellence
- Empathy/compassion
- Community
- Independence
- Positivity
- Motivation
- Respect
- Perseverance
- Flexibility
- Focus

- Curious
- Commitment
- Determination

If students feel valued, listened to, and respected, they will put their best foot forward to work toward a common goal.

YOUR ROADMAP FOR AN INSPIRING LEARNING ENVIRONMENT

From the outset, knowing your roadmap for the individual, for the class, and for society as a whole need to be a part of your game plan. As educators, we are charged with preparing children for success so that when they reach adulthood, they can be contributing members of their community. It's also important to have flexibility with the roads you choose because as you work through a school year, you may need to pivot the strategies you thought would work and adopt new ones as students' skills evolve. Following are some things to keep in mind when devising your roadmap.

Set Expectations

To perform well in a collaborative, problem-solving environment, kindness must come first. If students feel judged, ranked, or belittled in any way, they (of course) will be less willing to share and have the vulnerability necessary to take intellectual risks. The room setup needs to be such that it is conducive to active learning and students can easily interact with each other (tables versus rows of desks).

Letting students know your personal teaching philosophy as well as your overall goals is helpful. Again, they should be a part of the process so that you are working together as a unit and not as an authoritarian figure. If they are connected with the roadmap, they will have more of a vested interest in meeting goals. They need to understand that the class values kindness, learning from mistakes, taking risks, and positive collaboration. Create your mission statement and core values so that students can see these values every day and strive to meet them.

Know Your Students

Taking the time for the Finding Your Venn exercise or other methods of your choosing to get to know your students as individuals first and then as a class second will allow you to create a well-defined team of problem solvers. Today, schools and teachers tend to fall into one of two camps as either fans of the clumping method (having those with higher talent grouped together) or the scatter method (having students of varied skill levels mixed together). In the end, neither is as important as understanding what learning objective you have for the day and how the individuals in the class can maximize their understanding while working in a group. What is more critical than mathematical talent is students who can bring out the best in each other so that the journey is enriching.

That means walking a fine line between frustration and challenge. Problem solving should allow students to break free from the tyranny of 100%. If a student is understanding 100% of the material in the class, then they need more challenge. The aim would be for them to be in the 50–70% range so there is still challenge, but not frustration. Up to 50% is the foundational knowledge necessary to understand the material and stave off frustration. That is because it is about developing thinkers and not computers.

Have a Plan

Goals should be few and strategies should be abundant. In other words, when developing a roadmap for the school year, you want a few main goals, but the strategic initiatives to get there must be flexible enough to change with the altering landscape of your class and how it changes over time. This is the same type of flexibility required when solving a problem, so it only makes sense that is should be a part of the overall curriculum design. Feedback loops are essential so that teachers know what works and when to pivot.

Teaching is an iterative process, as is problem solving. You test strategies, take the best practices learned from those approaches, and iterate to the next level while keeping the objective for the lesson in mind. The best way to assess best practices is through feedback loops. Many times, students are nervous to provide quality feedback, especially if their name is attached to it. Having a communication channel that allows for anonymity allows students the freedom to provide constructive feedback.

At our AwesomeMath Summer Program, when utilizing this anonymous feedback channel, it was interesting to have three students request that scores not be given on their tests, but instead, they wanted more detailed feedback as to how they could improve each of their solutions. It was more work, but the class saw greater improvement over a shorter period of time.

Develop Your Style

Having knowledge of your own personal style when working with students helps to provide the insight of how you can best guide a classroom. A classic interview question when applying for a job is, "Where do you see yourself in the next five years?" The purpose of the question is to test the forward thinking of the applicant. As a teacher, you can have a new job every year as a new crop of students migrates through your class – every year, it is a new job. What would your forward-thinking question be? One that is asked of every AwesomeMath Summer Program applicant is, "What three adjectives would your campers use to describe your style?" It is an opportunity for the applicant to determine what value they want to add to the program based on their own personal drivers. This is also a technique when creating a resume or writing a cover letter. What three qualities of the applicant would shine through when read by the hiring manager?

Understanding how your teaching style engages kids in your classroom helps create a positive atmosphere conducive to learning. Learning from systems that have proven success, such as growth mindset, agile development, and design thinking, can inspire you to create an effective and student-centric classroom that allows each student to be their best version.

Having the ability to inspire your students to be better problem solvers will affect their lives for years to come and, in turn, they can aspire to rise to the example you have set for them.

Notes

1. Titu Andreescu and Branislav Kisacanin, *Math Leads for Mathletes*, Book 1, page 44, part of "Fun Sequences" (Providence, RI: American Mathematical Society, 2014).
2. Andy Ellwood, "The Dream Team: Hipster, Hacker, and Hustler," *Forbes* (August 22, 2012), http://www.forbes.com/sites/andyellwood/2012/08/22/the-dream-team-hipster-hacker-and-hustler/#54c48e312c85.
3. Titu Andreescu and Branislav Kisacanin, *Math Leads for Mathletes*, Book 2, page 14, part of Fun Sequences (Providence, RI: American Mathematical Society, 2018).
4. Purple Comet! Math Meet 2019 contest, https://purplecomet.org.
5. Carol S. Dweck, *Mindset: The New Psychology of Success* (New York: Ballentine, 2016).
6. L. Todd Rose, *How We Succeed in a World That Values Sameness* (London, Penguin UK, 2016).
7. Andreescu and Kisacanin, *Math Leads for Mathletes*, Book 2, page 26.

CHAPTER 4

What Is the Telos?

Telos is from the Greek and means purpose or goal. It's the root of *teleology,* or the study of purposiveness. One important goal of any educator is to tap the internal motivation of their students so that they can efficiently and effectively learn the material necessary during the school year.

Daniel Pink, who wrote the book *Drive*, states that there are three things that tap internal motivation:

1. Autonomy
2. Mastery
3. Purpose[1]

What better way to learn mathematical concepts then through a problem-solving-based curriculum? Students have the autonomy to work on challenges in a student-centric environment. Since the problems are scalable and process is the focus over grades, they can work toward mastery instead of other false metrics. And finally, the math is meaningful, with real-life problems giving them connection and purpose.

Autonomy to Solve Your Problems

Having the power to work on problems in your own way, trying different paths and approaches, coupled with kind guidance from the teacher, provides students the autonomy to learn at their own speed and by choosing their own paths. This autonomy gives them the *choice* to be their best version and develop understanding for where their particular set of skills adds the most value to various problems and various groups.

It's particularly empowering when students don't have to follow a prescribed set of steps to solve a problem but can *try* different approaches and *discover* what works. What would you prefer – to watch the slides from someone else's vacation and follow the exact same path OR learn some basics about the destination and then create a vacation plan that works best for you? We seek out recommendations from others, but in the end, the autonomy to choose and research on your own trip makes the vacation more customized and more interesting.

And, as always, how much people want to choose themselves versus what is chosen for them is a spectrum. You may have individuals who want to buy the package vacation versus those who want to wing it and see where the vacation takes them. It's the same in education, and as an educator, you can see where each student falls on this spectrum and help guide them in their journey.

This is where having feedback channels is so important. Whether they are anonymous or during the problem-solving session, teachers will have the opportunity to keep their finger on the pulse of the students' abilities and know when they may need extra time or support material to help with certain types of problems. In this interactive learning environment, students will also be working together, so not all of the explanations need to be made by the instructor. Many times, another student can explain concepts more clearly since they have more empathy with what it takes to learn the concept, having just learned it themselves, as well as being more relatable since they are closer in age to the student they are helping.

As you can see, problem-based learning puts a lot of stress on collaboration, and we can't overstate this enough: When you engage in the trade of ideas, everyone improves. So how does this fit with autonomy? The reason is because students still work at their own pace. With problem-based learning, you aren't lockstepping the kids together at the same rate; instead, you are letting them develop in their own way and at their own speed. Students are clever and they know when they are given the control to grow versus told how to grow. The autonomy to take the reins and guide their learning inspires them to excel. When expectations are high, students want to reach those expectations, and when they are low, they will do the bare minimum to check the box. By seeing their value and giving them autonomy, you raise the expectations and, in turn, show respect for their ability, which provides tremendous motivation.

Collaborative problem solving will allow kids to blossom into leaders, adept problem solvers, innovative thinkers, process managers, knowledge facilitators, and more! The main point is that they have the autonomy to grow authentically for who they are.

Mastery Through Inquiry

As stated earlier, when you shift the focus from outcomes (grades, rank, solutions) to the journey (thought process, multiple approaches, creativity), then students learn to work toward mastery of a topic and not GPA or other false metrics of ability or worth. When false metrics are removed, then students can be given problems that may normally be outside their grade level and take the intellectual risks to attempt the problem.

In mathematics, the art of proposing a question must be held of higher value than solving it.
Georg Cantor

Working toward mastery requires vulnerability, and that means asking for help. Being able to ask the right questions is a skill that lasts a lifetime. As an educator, it is imperative to create a culture of curiosity, and you can do that by training students to continually question. Unfortunately, in today's education system, students are expected to check the boxes of a rubric or standardized test rather than question and think outside the box. By taking risks and allowing for the vulnerability to make new discoveries, students can follow their curiosity, ask interesting questions, and work toward mastery.

At the AwesomeMath Summer Program, where students from all over the globe train to improve their competition mathematics skills, there will frequently be instructors in the lower level classes who will give students a problem(s) from the International Mathematics Olympiad (IMO) where only six students can earn the honor to represent their country at this annual event. While the problem will be from a time when the IMO was not as difficult as it is today, the problem will still be outside these student's skill levels, and yet, the synergy of working with peers under the guidance of an enthusiastic instructor will provide the confidence necessary to tackle a truly difficult problem so that they can increase their mastery of the subject. Students will roar with delight at their ability to compete at this level and be determined to try more problems in the future.

Keep in mind, however, that learning is messy and each student climbs the tree of knowledge in a different way. The best way to guide them on the climb so they can master that tree is by creating an environment where questions are welcome.

So How Do You Create a Space Where Students Can Be Vulnerable and Ask for Help?

Model the behavior. If students aren't asking questions, then lead with questions you think should be asked for a particular problem.

Encourage all questions. Praise students for asking all levels of questions, because if they are confused, chances are there is another student too shy to ask.

Have students roleplay asking questions. Make the time for this exercise in every class.

Be vulnerable and make mistakes in front of your students, asking them for help to solve the problem, which demonstrates it is about the process and not knowing everything from the start.

Be excited to share and go beyond the lesson. When your own curiosity is peaked, the students catch your enthusiasm and want to chase down the thought with you. Following the "what if" train is a lot of fun!

- **Build suspense.** Ask leading questions to a problem so that the process slowly unfolds in front of the students and they see the joy of figuring things out.
- **Have students collaborate** in small groups. They can guide each other to asking better questions.
- **Create feedback channels,** so students can ask you questions and follow their curiosity.
- **Determine where students struggle with asking questions;** is it language, critical thinking, inhibition, etc.? Then, you can provide question frameworks to get them started on the path of inquiry.
- **Provide question feedback.** In language classes, students are able to peer review each other's work. Bring this to the math class and create a document where students can ask questions that can either be critiqued by you or other students. It should *always* be constructive feedback and help shape the questions to get the most from the answer.
- **Bring back the fun.** Start playing with ideas, manipulatives, history, etc., so that students open up and are more willing to question and extend ideas.

These ideas are the tip of the iceberg. When you start working with your students to create an environment where questions are a key component, then their curiosity will be unleashed, and they can begin to master the topics to which they put their energy, play, and thinking.

Purpose with Competitions

Problem solving provides purpose through meaningful and relevant challenges. The competition aspect, while not the end goal, is a wonderful way to provide connection to a larger community. In fact, our preferred term is the portmanteau of cooperation and competition, *coopertition*. Students work together to solve problems, not so they can be better than their classmates, rather, so all of them can improve and celebrate each other's progress.

The math competition community is a kind and supportive group where students and teachers alike enjoy discussing problems. There are numerous competitions with various styles, structures, and goals, however, the one common thread is that they are created by mathematicians who want to excite and engage students in the joys of solving interesting problems.

Just like other competitions, such as sports, there are benefits when working toward common goals in a teamlike atmosphere that provides all team members with purpose and connection. Purpose is being a part of something larger than yourself, and the connection is your common goals. Competitions provide the components for the math team to have purpose and success by providing:

- Clarity about roles played by members of the team
- Agreed training processes

- Mutual mission and core values
- Expectations for kindness in all interactions
- Feedback loops

Being able to work on teams enhances:

- Efficiencies
- Focus
- Creativity
- Risk-taking
- Support
- Trust
- And more...

For a team to be effective, the focus must be on the goals or the problem, and not the individual. You want to nurture the participants so they feel a part of the team, but because you are competing for improvement (not necessarily winning, although that's nice, too), the objective is for the team to grow in ability and then the individuals will, in turn, benefit. It's about working together, not ranking, so that every member can grow and find their purpose in the group.

When coaching math teams, one way Kathy set up the room was to have students at various "level" tables. Because she didn't want them to rank each other by being at Level 1, 2, or 3, instead, they would be seated at tables labeled as famous composers and the rank was based on when that composer was born. The composers, in order of birth year, were Bach (born in the year 1685), Mozart (born in the year 1756), and Beethoven (born in the year 1770). If the lesson for the day was on concepts in algebra, students who have that strength would sit at the Bach table and those who may need more guidance would sit at the Beethoven table (they would self-select). Everyone was at a well-renowned composer table, but their table was determined based on experience as it relates to the composer's birth year, nothing more. If you were at the Beethoven table and had a question on how to solve a problem, you would ask someone at your table and/or someone at the Mozart table. If you had a question at the Mozart table, you would ask someone at your table and/or someone at the Bach table, and students at the Bach table could confer with each other or with Kathy as their coach.

Students are remarkably good at knowing their own strengths, and it was never an issue for them to select a difficulty level, especially when given a handout of the problems at the beginning of class. Sometimes they would ask for advice, and then Kathy would give them guidance as to what would be the best fit. Again, when the curriculum is about the information and not about judging the student personally, then there is less posturing and concern over place. They aren't being graded; they are pursuing a worthy task, and that journey provides *purpose* as does the community in which the discoveries take place.

Quadrants of Success

As an educator, one important goal is guiding students toward success, but what does that look like? It would make life much easier if there was one formula for success that everyone could follow, but this does not exist. Students can't follow a program en masse and then all will be well.

Further, success means different things to different people. Luckily, there are some common factors that play a role in the success of every student, regardless of where their life journey takes them, e.g., mathematician, dancer, engineer, artist, biologist, linguist, computer science, or any combination of areas!

These four quadrants apply when pursuing problem solving and other areas of life as well. Your talents, choices, people, and timing all play a role in growth and development.

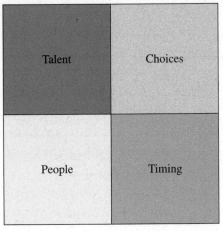

TALENT

As with all things, talent matters. For example, a student can train for 15 hours a day but still not be an Olympic gymnast if they don't have the flexibility or body type for this sport. They could, however, learn to be physically fit and embark on some gymnastic moves that fit their body and their interests. The same is true with mathematics, everyone can improve and get better by understanding *their* particular talents and then determining what areas are the best fit. Some may want to compete in mathematical Olympiads, but others may have talents better suited for mathematical research. Still others may only wish to acquire the critical thinking a problem-solving curriculum can provide to pursue careers outside of mathematics.

Once students discover where they shine, they can nurture that talent and curiosity and take the topic to greater depth. Mathematics has so much depth and breadth, **all** students can find an area that showcases their talents. Don't confuse talent with being good at math. Mathematics is too broad a subject for talent to be a binary choice – good or bad. It would be like saying you don't like reading when you've only read one style of book. Everyone is good at math; however, their talent will come through in some areas more than others.

Because a problem-based curriculum allows students to move beyond the regular subjects, such as algebra and geometry, and also explore discrete mathematics topics, logic, and more, they can see what ignites their curiosity and inspires them to deepen their understanding.

It's when a ceiling is placed on exploring talent that a student can become bored or disengaged. A problem-based curriculum, which is scalable, removes those ceilings and opens the possibilities for each student to study a topic to its edges and beyond.

CHOICES

Everyone is a product of their choices. A question that comes up all the time from parents is *"How do I limit my child's time playing video games?"* Time is such an arbitrary way to affect choices. How much is enough time to practice piano per week to see progress? How much time is too much to spend in one area? How long do you need to be around peers to count as socialization? The answers depend on the person, their goals, their abilities, etc.

Malcolm Gladwell became famous (or infamous in some circles) with his 10 000-hour rule as stated in his book *Outliers*.[2]

Ten thousand hours is the magic number of greatness.

Malcom Gladwell

To attain elite status and truly be good at something, Gladwell contends that you must practice early and often so that you can hit the magic "10 000 hours" number. But that isn't the full story – it has to be an area in which you are passionate and talented – and most students won't take the time to even play video games for 10 000 hours unless they are the small handful who want to compete at a professional level.

The answer to guiding a child to make good choices is by, again, giving them control (autonomy) over their outcomes. The parent or teacher can ask their child a simple question: *"How much time do you spend consuming things other people have made and how much time do you spend creating something original?"* That said, you can actually consume very worthwhile things, for example, education. Students can also creatively produce a lot of things that don't necessarily have value. So an extra caveat must be added of active versus passive. Consuming with purpose (active) is a lot different than passive consumption.

In mathematics, are you *actively consuming* interesting problems with the goals of:

- Testing your skill sets in math competitions?
- Reaching deeper understanding in a topic?
- Learning how to apply mathematics to different areas?
- Understanding the world better?
- Progressing in school?
- Challenging yourself?
- And so, on....

When you are producing content and creating new ideas, who is your audience? Humans don't just produce solely for their own purposes; they usually want to share with the community. Granted, the community with which information is shared can be very small, but no man is an island, and most ideas or items are produced with the intention of adding value to a community in some way.

With a problem-based curriculum, this type of *active creativity* or production can take the form of:

- Learning to write your own problems
- Creating a curriculum
- Exploring mathematics to its edges (pure math)
- Depth and breadth of topics outside the standard curriculum

Having a balance between active consumption and active production so as to learn and grow every day is a worthy goal for every human being. In mathematics education, it provides students with both the short view, what they need to accomplish to be successful in school, and the long view, what they need to learn to be successful at whatever occupation they choose.

PEOPLE

Some People
by Rachel Field

Isn't it strange some people make
You feel so tired inside,
Your thoughts begin to shrivel up
Like leaves all brown and dried!
But when you're with some other ones,
It's stranger still to find
Your thoughts as thick as fireflies
All shiny in your mind!

People have the power to bring out the best in us and the worst. The same is true of our environments. If you create an environment where the expectations are set ahead of time for kind, positive guidance and collaboration, then the result is a room full of firefly thoughts and not brown, dried leaves. A large factor in the success of individuals is the people with whom they surround themselves.

How Have Your Teaching Methods Evolved Over Time, and Why?

Dr. Branislav Kisačanin: The social aspect of collaborative problem solving is immeasurably beneficial for developing young minds. You cannot put a price or cost/benefit number on:

1. Knowing that they are not alone in their love of knowledge and curiosity.
2. Positive peer pressure (yes, there is such a thing, even though we only hear about the negative peer pressure).
3. Making life-long friends they will encounter again in college, scientific conferences, and when they travel to Stockholm to pick up their Nobel Prize.

As a mathematics teacher, you want to ensure that each student is focusing on the problems at hand and not external things that can be divisive or based on status. If the room is focused on truth, e.g., trying to solve engaging problems, and not power, e.g., who is best, then they can work together in a productive way.

The distinction between a truth versus power approach is an important one. Today's positive shifts in education with collaborative problem solving, growth mindsets, and design thinking, as mentioned before, are student-centric for the purpose of inspiring students and defocusing outcomes. It's critical that students are surrounded by people who want to see them improve and motivate with positive energy. These methods are how you employ a truth approach. Power approaches involve focus on outcomes, egos, shaming, and comparing.

Students, when the expectations are raised within a supportive environment of kind peers, will live up to their better selves and their peer relationships will strengthen. Working with friends to solve interesting problems forges friendships that go well beyond high school. Many of our AwesomeMath Summer Program students, who meet for the first time during the three-week program, stay in contact throughout high school, college, and beyond and will still share ideas and fun math problems with each other. Further, the memories and relationships are so strong that many of the residential and teaching assistants are former campers who return to the program so they can continue to be a part of this kind, engaging environment.

The people you choose to surround yourself with have a lasting impact, so to be successful, it is critical to be a part of a community that values your contribution and, in turn, you value theirs.

TIMING (LUCK)

Timing has always been a key element in my life. I have been blessed to have been in the right place at the right time.

Astronaut Buzz Aldrin

Yes, an entire quadrant is devoted to luck, timing, and good ole serendipity. Sometimes, they are in your favor, and sometimes they are not.

There are many times when an individual will have a great idea, but:

- The world isn't ready for it.
- The idea is incomplete.
- The execution of the idea misses the mark.
- Or something else entirely creates a roadblock.

In the late 1600s, Isaac Newton wrote a letter to fellow English scientist Robert Hooke, in which he famously said, "If I have seen further, it is by standing on the shoulders of giants." Newton recognized that his major breakthroughs were only possible because of the groundwork laid by others, and many times, the world would not have been ready for his ideas if they had happened any earlier.

The world changes at such a pace that if individuals and organizations don't know how to think and pivot during their journeys, they will be left behind. That is why problem-solving-based curriculums manage both the short and long view of mathematics education – one foot must be firmly planted in learning in the here and now while the other steps forward toward the future and how to apply the knowledge gained. In this way, luck can be better managed by keeping opportunities open and putting forth the effort to mitigate its negative effects.

Problem-based learning depends heavily on feedback channels so that ideas can be tested, what works is moved forward, areas that need more development are refined, the process iterates, and the individual moves closer to achieving his or her goals.

Each of the success quadrants are interdependent; therefore, timing and luck require the individual to be a part of community (people) so that feedback and guidance can be available. Then there must be the ability to choose a path and have the talent and skills to make a difference.

Notes

1. Daniel H. Pink, *Drive: The Surprising Truth about What Motivates Us* (New York: Riverhead Books, 2009).
2. Malcolm Gladwell, *Outliers: The Story of Success* (London: Hachette UK, 2008).

CHAPTER 5

Gains and Pains with a Problem-Based Curriculum

There are no perfect systems; therefore, it is important to analyze the value proposition of a problem-based curriculum with regards to the *pains* and *gains* of the participants, namely:

- Teachers
- Students
- Parents

Your value proposition chart may not match those below; however, each chart should be viewed as a fluid list that you can customize based on your requirements and situation. To have learning and growth, the *gains* need to be significant and there should be *pain relievers* for the potential negatives. When you know each participant's *pains* you can:

- Pivot within the system.
- Employ resources for the system.
- Analyze the variables that may affect the system.

Based on discussions with teachers, parents, and students, these are a few gains and pains charts for a problem-based approach in the classroom. You'll note that some of the pains stay the same for each group, but this is especially so for the gains since in the end, students, teachers, and parents alike are seeking a good education.

Let's look at one example from the teacher's perspective, where a pain can be relieved by a pivot, resources, or analysis. For example, there can be a time constraint that diminishes the effectiveness of a problem-based curriculum. *When a class is only 45 minutes long, how can you introduce a new concept, e.g., divisibility rules, and still have meaningful collaboration amongst students for how to solve problems related to the concept?*

- A pivot could be to provide them a list of divisibility rules to review for homework with some example problems and then provide the challenging problems in class.
- Resources could be to create a flipped classroom video of the lecture to be viewed at home and the problems can then be the focus of the class.
- Analysis of, for example, classroom management can help to free up time that is wasted in class to get the most out of each moment and effectively learn the lesson in conjunction with the collaborative problem-solving component.

Teachers

In today's world of high-stakes standardized tests, the fervor that students should all go to college, and pressures for school funding, this means that teachers don't always have the flexibility to teach beyond the rubric and are bound by the pacing requirements (scheme of work) and standardized test pressures set by the district, state, or country in which they work.

Being a teacher today means working inside a potentially inflexible structure of top-down directives, exam boards, and standardized tests. Students' skill levels can vary greatly, depending on their own learning speeds, family involvement, and/or educational backgrounds, making it more difficult than ever to manage the metrics to which each teacher is held accountable. On top of educating each individual student to his or her needs, teachers also need to have efficient classroom management, time to plan lessons, and interact continuously with parents, making the hours of the day seem very short indeed! The intent of this book is to provide the scalable curriculum and lessons that will reach a wide array of levels and abilities so that the teacher doesn't have to expend valuable resources, time, and energy to create the material from scratch. Understanding the obstacles (pains) and rewards (gains) from a teacher perspective will allow the value proposition of problem-based learning to come to fruition.

There will be differences in the gains and pains, of course, which are location dependent (city, state, country) and type of school dependent (public, private, homeschool). That is why this is not just a declaration of perspective each teacher will have, but instead, a starting point to get the educator to think about the value a problem-based curriculum can provide for their student.

Value Proposition: Teacher Perspective

Gains

- Teacher Efficacy
- Deeper Learning
- Retention
- Student-Centric
- Active Engagement
- Interpersonal Skills
- Meaningful Problems
- Continuous Challenge
- Test Scores

Pains

- Time Intensive
- Student Skill Gaps
- Teacher Inexperience
- Pacing Requirements
- Standardized Tests
- Class Size
- Class Duration
- Efficiencies
- Incomplete Curriculums

TEACHER GAINS

Teacher efficacy	When using a problem-based approach, teachers have noted improvement in their own mathematics skills and teaching abilities.
Deeper learning	Their students, when made to think critically, learn deeper concepts than they would in a broadcast environment.
Retention	When tested later, student's retention of topics lasted longer with a problem-based approach. This is because they are no longer just memorizing concepts, they are developing them and so they are able to reconstruct what they've learned and/or derive formulas themselves, because they understand how they were created.
Student-centric	Students who are at the center of their own learning experience are more willing to participate and see the value in the lesson.
Active learning	Problem-based learning, because collaboration is such a large part, creates an active learning environment where students ask questions and work willingly on the problem sets with peers and their learning facilitator (the teacher).
Interpersonal skills	Conflict resolution, problem solving, decision making, and leadership skills all flourish in a team-based collaborative environment.

Meaningful problems	When the problems connect to real life for the students and they see the use beyond the class, they are more curious about the material and intrinsically motivated.
Continuous challenge	A problem-based curriculum provides challenge levels for all the students in the class (scalable) and since the focus is on process and not outcome, problems can be harder.
Test scores	Teachers noted that while the problem-based process takes longer, the payoffs are greater since higher retention and deeper learning lead to the critical thinking required to succeed on standardized tests and beyond.

TEACHER PAINS

Time intensive	Carving out collaborative problem-solving time into a single class period along with a short lecture of the topic is difficult. Plus, since many of the problems are multi-step and can be approached in different ways, there is more upfront prep time required and more time required on the part of the students to think through the solutions.
Student skill gaps	There can be a wide gap in skills between students in the class, making it difficult to find a level where kids who need more challenge receive it and students who are missing foundational understanding are brought up to speed.
Teacher inexperience	The teacher may not have the mathematics background necessary to teach problem solving at this level.
Pacing requirements	The school, district, and state may have pacing requirements, which dictate a demanding schedule of topics, making a broadcast approach more expedient even if it isn't more effective.
Standardized tests/exam boards	Some schools, states, and countries have required testing for all students where certain scores must be met to receive funding or other benefits, shifting the teaching incentive away from challenging mathematics and toward teaching to the test.
Class size	Classes over 20 students present a challenge to teaching a problem-based curriculum, especially if the teacher is inexperienced in classroom management.
Class duration	A 30- or 45-minute class period can create a constraint to problem-based learning, especially if it is a large class.

Efficiencies	A student-centric approach lacks the efficiencies of a teacher-centric broadcast approach, especially when transmitting information to a large group. However, the broadcast approach is based on the fallacy that the communication is being received and understood.
Incomplete curriculums	Many of the problem-based materials supplied to teachers today are incomplete and don't provide the necessary support to make this approach attractive to use.

How can these pains (obstacles) be overcome to make problem-based learning viable in today's schools? In Section II, Teaching Problem Solving, you will have access to curriculum for a full class period, 45–55 minutes, as well as *mini-units* (Chapter 8), 10–15 minute exercises comprised of *hook problems,* which are intended to get the kids thinking deeply while being scalable to various skill levels. The full-unit material provided in Section III, as well as the mini-units, will provide not only lessons, problems, and solutions, but also leading questions that the instructor can ask to ensure the students are driving the process instead of rushing to the outcome. This approach will aid teachers in keeping the pulse on learning for each student and the class as a whole, making student assessments easier and more streamlined.

By providing the curriculum, questions, techniques, and solutions for every educator, the pains detailed above can be relieved and, in some cases, cured. There will still be the need for proper planning and classroom management skills; however, the mini-units are a great way to get started with a problem-based approach as well as gain the experience necessary to bring this effective method to life in the classroom.

Further, it takes time to implement a problem learning curriculum, but the time required can be minimized if the teacher is a part of a professional learning community where ideas can be shared and strategies discussed. This community is essential to streamline the process and gain the efficiencies realized by working with like-minded peers.

Students

The amount of free time a student has today is significantly less than it used to be. They are overscheduled and have less freedom to play and explore. Between after-school programs, sports, enrichment activities, family obligations, and schoolwork, it is difficult for them to have the brain space to focus on hard problems in a significant way. Add in lack of sleep due to juggling so many commitments and it is even more difficult to be connected and engaged.

Further, for many students, *math* is a negative word, and they can easily fall prey to either:

- "I'm not good at math."
- "When will I ever need this?"

This is especially true in cultures where it's okay to completely discount the importance of mathematics, and yet, so much of their future depends on mathematical literacy. It is in the student's best interest to engage in due diligence and look at the myriad career options that are available to them. Every aspect of their lives will be touched by mathematics, from personal finance to big data to just understanding statistics in news reports, and both their careers and personal lives will require some level of mathematical competence.

Problem solving brings real-world problems to their front step and shows them how imperative it is to have mathematical understanding. Having accessible manipulatives, hands-on projects, and real-world circumstances as a part of the problem-solving process will make the material relevant, authentic, and useful – this is more than can be said for video games!

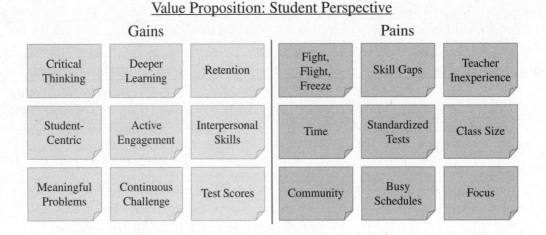

STUDENT GAINS

Critical thinking	Multi-step problems that can be solved in more than one way train students to think critically in order to work them out.
Deeper learning	When made to think critically, they learn concepts at a deeper level than they would in a broadcast environment.
Retention	Retention is much greater when working at a more challenging level and engaged in the resilience of problem solving.
Student-centric	When you are at the center of your own learning path, a student will have more accountability in the process and take more ownership of the outcomes.
Active engagement	Collaboration with peers and educators makes for a more rewarding experience and active environment.
Interpersonal skills	Conflict resolution, problem solving, decision making, and leadership skills all flourish in a team-based collaborative environment.

Meaningful problems	When the problems connect to real life for the students and they see the use beyond the class, they are more curious about the material.
Continuous challenge	A problem-based curriculum provides challenge levels for all the students in the class, and since the focus is on process and not outcome, problems can be harder.
Test scores	Knowing how to approach problems, deconstruct them, think critically, and imbibe foundational material increases student test scores.

STUDENT PAINS

Fight, flight, freeze	Fear of mathematics and "appearing stupid" causes students to have a "fight, flight, or freeze" response that they need to learn to work through.
Skill gaps	If students have gaps in the foundational material necessary for the class, they can feel behind and, in turn, disengage.
Teacher inexperience	The teacher may not have the mathematics background necessary to teach problem solving at this level.
Time	Some students require more time to process and absorb information, so the time-intensive nature of solving difficult problems can be an obstacle for them.
Standardized tests/exam boards	Having to focus on standardized testing requirements for a large part of the school year can demotivate and distract students from investing time in problem solving.
Class size	Classes over 20 students create a chaotic environment where it can be difficult to focus if the teacher cannot manage the class.
Community	With busy after-school schedules, students may not have the time or ability to reach out to their peers and community for guidance and motivation.
Busy schedules	Problem-based learning requires slow and deep thinking for which today's overscheduled student doesn't have the time.
Focus	So many students today are diagnosed with some sort of attention deficit issue or may not be getting the time during the school day to burn excess energy, which causes problems with focus in large classes with lots of distractions.

One significant pain faced by students is math anxiety (i.e., fight, flight, freeze), and as is true with all anxiety, what you are really trying to combat is fear. So how do you help students (and, in turn, how they can help themselves) get over this fear of math? One way is to understand its origins. Math fear can be transmitted from well-intentioned parents, friends, or teachers, as well as originate from other more personal areas such as the following:

Perfectionism. Not wanting to make mistakes.

High standards. This is slightly different from perfectionism in that it's not fear of mistakes as much as holding yourself to high personal expectations of performance, and if you don't think you can meet them, you won't try.

Social awareness/empathy. This is when you have an understanding of your classmates and their abilities and don't want to appear stupid by comparison.

Imposter syndrome. When you discount the abilities you do have as just luck or a fluke and believe that you really aren't good at math, just situationally serendipitous.

There are steps an educator can take to help students face the fear of mathematics:

1. *Name the fear.* Try and understand the origin of their feelings so that they can move from irrational feelings to a more rational approach.
2. *De-catastrophize.* Students will jump to the worst conclusion when things are scary. Help them to understand what they know about their fear and what they can control so they can take charge of the problem instead of their anxiety being in the driver's seat.
3. *Cope*: Develop a plan for how to handle math phobia in the future.

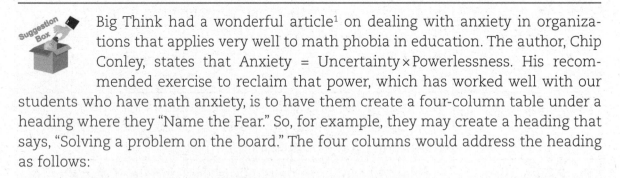

Big Think had a wonderful article[1] on dealing with anxiety in organizations that applies very well to math phobia in education. The author, Chip Conley, states that Anxiety = Uncertainty × Powerlessness. His recommended exercise to reclaim that power, which has worked well with our students who have math anxiety, is to have them create a four-column table under a heading where they "Name the Fear." So, for example, they may create a heading that says, "Solving a problem on the board." The four columns would address the heading as follows:

1. What do you *know* about this fear?
2. What don't you know about the fear?
3. What can you control about the fear?
4. What can't you control about the fear?

Columns 1 and 3 should be much longer than columns 2 and 4. That is because the things you know and the things you can control are in your power to change. The thing you don't know and can't control is the student catastrophizing about possible outcomes that aren't rational. Following is a sample table answered by a student.

SOLVING A PROBLEM ON THE BOARD IN CLASS

What do you know?	What don't you know?	What can you control?	What can't you control?
The problem to solve.	You don't know if you'll get the correct solution.	You can practice presenting problems to smaller groups.	You can't control the reactions of others or what they think.
The strategy to solve the problem.	You don't know if you'll freeze when trying to solve the problem.	You can work to master the material and try your best.	
You've been given the mathematical understanding for how to solve these types of problems.		You can talk with the teacher or peers if you need more guidance.	
Your instructor and peers are able to help if you get stuck and the environment is supportive and kind.		You can present problems to the class that are easier for you to solve so that you gain confidence for harder challenges.	
There are no negative consequences if you get the problem wrong.		You can make sure you take care of your physical needs (sleep and exercise) so that you are mentally ready for the challenge.	
You've solved similar problems before.		You can practice at home in a less stressful environment.	

Further, if students have skill gaps, meaning that they are missing the foundational knowledge to make progress in a curriculum unit, this can feed their math phobia. This is why a curriculum that has scalable material for all skill levels introduced in an environment where mistakes are valued as opportunities for growth makes all the difference.

If you have students who receive quite a bit of supplemental enrichment in problem solving, their skills may surpass the experience of the teacher. At the AwesomeMath Summer Program, we encounter these exceedingly bright kids all the time and have the faculty necessary to either increase their class level or pivot the class material quickly to reach the needs of even the most gifted problem solver. Not all schools have this luxury, and many times, truly gifted mathematicians are left to fend for themselves. This is why the curriculum used in the school is so important. Even if a teacher has skill gaps of their own, they can easily follow the teacher portion of the unit lessons with the following beneficial additions:

- Thought-provoking questions to guide all levels of thinkers
- Scalable material that shows breadth and depth of concepts
- Challenge problems so that even a gifted learner will be excited by the material

These additions will allow students of varied skill levels to connect with the material while also providing enough range that students can work at their own processing speed. That said, no curriculum can be perfect, but if the material is explained well, interesting, and scaffolded, then students will be able to learn and grow.

A difficult set of pains to overcome are community and busy schedules. What truly makes a problem-based learning curriculum come to life is when students organically form their own problem-solving groups and collaborate. Just like in gaming, when you get online with friends to play, it creates more excitement and connection with the material when you can share the experience. However, after-school schedules are so jam packed that this time does not exist. Some schools are trying to combat the issue by creating an *open period* during the school day where students can talk about homework and work on projects, but this needs to be well managed to work effectively, and the best way to do that is to identify a few students to take on the role of group leader and cheerleader for the material.

 When coaching mathletes, Kathy recognized that some students are *sensory seeking*, i.e., they crave a lot of input and connections, while others are *sensory avoiding*, i.e., they need to isolate and block input so they can look into their mind's-eye and think. To accommodate both types of students, she kept a box of fidget items that didn't make sound as well as allowed *movement thinkers* to pace in the back of the classroom and *reflective "mind's eye" thinkers* to sit in the front of the room where it was quieter.

Another significant consideration is distractions that can happen externally such as the added chaos of large class sizes or internally with issues such as attention-deficit hyperactive disorder or sensory sensitivities. A benefit of a problem-based learning model is that it doesn't limit movement; many students process information by walking around the room or fidgeting, and it's important to let this happen as long as it doesn't distract others.

Parents

Today's parents are wearing a number of different hats with work, raising kid(s), extracurricular activities, after-school programs, and balancing their own lives. They rely on the schools to prepare their child for what comes next in school, what they need to be successful in college, and/or for the workforce they will be entering.

Families are complicated and come in many different forms:

- Various marital statuses such as divorced dual custody, single-parent households, married, remarried parents with other children, etc.
- Various employment combinations such as both parents working, single income, part-time, work from home, etc.
- Various education backgrounds and educational expectations
- Various multigenerational families living together
- And so many more combinations

All these types of families mean that the pains and gains from a parent perspective with a problem-based curriculum can be just as complex as the lives of the parents.

Value Proposition: Parent Perspective

Gains			Pains		
Critical Thinking	Deeper Learning	Retention	New Math	Skill Gaps	Teacher Inexperience
Student-Centric	Active Engagement	Interpersonal Skills	Time	Standardized Tests	Support
Meaningful Problems	Continuous Challenge	Test Scores	Material Covered	Busy Schedules	Grades

PARENT GAINS

Critical thinking	Multistep problems that can be solved in more than one way train students to think critically in order to work them out.
Deeper learning	When made to think critically, they learn concepts at a deeper level than they would in a broadcast environment.
Retention	Retention is much greater when working at a more challenging level and engaging in the resilience of problem solving.
Student-centric	When you are at the center of your own learning path, a student will have more accountability in the process and take more ownership of the outcomes.
Active engagement	Collaboration with peers and educators makes for a more rewarding experience and active environment.
Interpersonal skills	Conflict resolution, problem solving, decision making, and leadership skills all flourish in a team-based collaborative environment.
Meaningful problems	When the problems connect to real life for the students and they see the use beyond the class, they are more curious about the material.
Continuous challenge	A problem-based curriculum provides challenge levels for all the students in the class, and since the focus is on process and not outcome, problems can be harder.
Test scores	Knowing how to approach problems, deconstruct them, think critically, and imbibe in foundational material increases student test scores.

PARENT PAINS

New math	The mathematics the student brings home may be completely foreign to some families, especially areas such as Number Theory and Combinatorics (Discrete Mathematics), making support with homework difficult.
Skill gaps	If students have gaps in the foundational material necessary for the class, they can feel behind and, in turn, disengage. This can be a negative for parents who want their student to be fluent in mathematical literacy.

Teacher inexperience	The teacher may not have the mathematics background necessary to teach problem solving at this level. Therefore, parents may need to step in for support, but not all will feel comfortable doing so.
Time	Some students require more time to process and absorb information, so the time-intensive nature of solving difficult problems can be an obstacle. If they need to work on the problems at home, there may not be enough time/support to think deeply.
Standardized tests	Having to focus on standardized testing requirements for a large part of the school year can demotivate and distract students from investing time in problem solving.
Support	If parents don't understand the importance or necessity of a problem-based curriculum, they may feel the effort is a waste of their student's time.
Material covered	The number of concepts that can be covered in a problem-based curriculum, due to the time required, may be less than what can be covered in a broadcast curriculum. That said, retention, understanding, and learning are deeper, but some parents may feel uncomfortable with the pace.
Busy schedules	Problem-based learning requires slow and deep thinking for which today's overscheduled student doesn't have the time nor do many families.
Grades	With problem-based learning, it is better to work toward mastery as opposed to grades, but in today's *check the box world*, out-of-the-box thinkers may not be as valued.

A problem-based curriculum can be a boon for parents since their children will be better prepared for the remainder of their middle and high school years as well as gain the thinking skills necessary to excel in college and jobs. When their thinking improves, so does their grade point average (GPA), test scores, confidence, and options. This doesn't, however, mean there aren't pains to providing the at-home support necessary to embrace the problem-based approach.

Parents who haven't been exposed to either discrete mathematics topics or a student-centric approach to problem solving may feel ill-equipped to provide the support necessary for their student. This is *new math* for them and can at first seem daunting. In our experience at AwesomeMath, many parents are seeking a *map of engagement*, in other words, a way to work with their child at home so they get the most out of their education. A map of engagement that will help a student truly excel in their education is one where parents guide their students on how to ask the right questions about a problem(s). Even though the math itself may be new to the parent, the skill of knowing *what* to ask cannot be emphasized enough, especially in a problem-based learning environment. Just as a consultant may not know a business as well as the company, consultants are still hired because they offer a fresh perspective and have the experience and knowledge to ask the right questions so the company can grow. Parents can be the at-home consultant who, through thoughtful questions, can help the child grow into a great problem solver. Guiding your child to ask good questions helps:

- Emphasize the process over outcomes.
- Focus their efforts to be more effective problem solvers.
- Show your vulnerability with problem-solving while at the same time your interest in learning with your child.
- Develop their ability to ask targeted questions that will translate to other areas of life.
- Strengthen their relationship with the teacher and add value to the class.

The curriculum in this book provides hints to problems that teachers can share with parents to guide them on asking questions, which is especially helpful since the math may be completely new to the parent. When a student works at home to write one or two questions to ask in class, they have more confidence and have thought more deeply about the material.

A math assessment test can help teachers determine any skill gaps a student may have so that they can be addressed and the gaps filled. This may mean providing outside resources such as online classes, videos, books, and/or tutoring. For example, some students in middle and high school may still struggle with understanding fractions that would be a critical gap to have assessed.

In some cases, especially parents who are already in science, technology, engineering, mathematics (STEM) fields, the parent may actually have more discrete math training than the teacher. This is why a curriculum that provides not only challenging problems for the students but also support materials for the teacher is so critical.

And as was discussed in the "Pains" sections for both teachers and students, time, busy schedules, and the emphasis on standardized testing can make it difficult to effectively implement a problem-based curriculum. Even if the curriculum can't take the place of the requirements dictated by the state or country and/or the students themselves don't have the time in their busy schedules to truly dive into a deep problem-solving approach, there are still mini-units available in the next section of this book where families can be exposed to the joys of problem solving, and hopefully that seed will have the time to bloom later on when external pressures and lack of time aren't so prevalent.

The final pain listed for parents is the importance of grades. GPA is a leading factor for college admissions, and so parents are less willing to give problem solving the time and energy necessary, especially if they feel that grades may be affected. As mentioned in previous sections, this is short-sighted thinking when raising life-long learners. How can you raise out-of-the-box thinkers in such a check-the-box world? Parents would love nothing more than to have students who are curious go-getters who can support themselves and be able to think, and yet, we all know the stress and pressure put on kids to check that box of GPA and standardized tests or exam boards. Unfortunately, this focus on a short-term objective has long-term consequences when 30% of all college freshman drop out.[2] Students who are still in middle and high school need to hit roadblocks in their learning, understand how to seek help and ask questions, and above all, know how to solve hard problems while they are still home and have support. Once they are in college, if they don't have these skills, it will be too late. Freedom to learn, think, and struggle in a supportive environment is what makes the difference between a check-the-box and out-of-the-box thinker.

Notes

1. Chip Conley, "Anxiety = Uncertainty × Powerlessness," Bit Think (May 22, 2013), https://bigthink.com/in-their-own-words/anxiety-uncertainty-x-powerlessness?fbclid=IwAR3_Kok8po2VH3OailbdlkdJ6eXnz43FgZnb1mBmctzg7jrha5baO-53UYE.
2. "US College Dropout Rates and Dropout Statistics," College Atlas (June 29, 2018), http://www.collegeatlas.org/college-dropout.html.

SECTION II

Teaching Problem Solving

In this section, we will provide teachers with the strategies and tools for guiding innovative problem solvers who can work independently and think deeply. And, as you may have guessed, this won't require complete conformity (e.g., "do exactly as I say") to a program. Rather, we encourage you to expand on the ideas presented to create a curriculum that works for your style and the students who make up your classes. Teaching is all about having a plan; however, anyone who has taught knows that lessons rarely go exactly as intended. Flexibility and creativity are required when working with students of varied levels and abilities, so this section seeks to provide a well-conceived plan, along with ideas from our teaching and learning community to inspire new ideas and methods. As detailed in Section 1, it's critical to create a kind and supportive environment that respects the intellectual abilities of all participants along with interesting problems, like the one below, that are scalable for students of different mathematical skill levels.

We have a large pile of identical 20 gram 1×1 square pieces. For some positive integers m and n the following displays use the same number of pieces:

- an $m \times (n+2)$ rectangle with a 8×8 square removed from its interior;
- an $(m-2) \times n$ rectangle with a 4×4 square removed from its interior;
- an $(m+4) \times (n-6)$ rectangle with a 2×2 square removed from its interior.

Prove that each of the three displays weighs 2020 g.

Solutions
Because the first two displays use the same number of pieces, we have $m(n+2) - 8^2 = (m-2)n - 4^2$, yielding $m = 24 - n$. Then, because the last two displays use the same number of pieces, we have

$$(m-2)n - 4^2 = (m+4)(n-6) - 2^2$$
$$((24-n)-2)n - 4^2 = ((24-n)+4)(n-6) - 2^2$$

In other words, we have $m(n+2) - 8^2 = (m-2)n - 4^2$, yielding $m = 24 - n$. Then $(24-n-2)n - 4^2 = (24-n+4)(n-6)$, yielding $n = 13$ and $m = 11$. Hence, each display has 101 pieces and weighs 2020 g.

This section is about helping educators be their better selves, as well and providing the information necessary to successfully launch a problem-based curriculum.

You will be provided with:

- Five Steps to Problem-Based Learning
- The Three Cs: Competitions, Collaboration, Community
- Mini-Units

CHAPTER 6

Five Steps to Problem-Based Learning

When creating a problem-based learning (PBL) program you need the problems, tools, management, and support to effectively guide each student toward the primary goal of creating innovative problem solvers. With the five steps outlined below, you will be well on your way to launch PBL in your classroom, utilizing the provided set of mini-units (10–15-minute lessons) in Chapter 8 or full units (45–55-minute lessons) in Section III.

1. Start with meaningful problems.
2. Utilize teacher resources.
3. Provide an active learning environment.
4. Understand the value of mistakes.
5. Recognize that *everyone* is good at math!

Start with Meaningful Problems

In this section, we will cover the following:

- What are good problems?
- Show relentless curiosity.
- Don't forget grace.

WHAT ARE GOOD PROBLEMS?

This book provides problems and lessons for your classroom; however, there is great joy in crafting and writing your own problems. We have all heard the expression, "Give a man a fish and you feed him for a day. Teach him how to fish and you feed him for a lifetime" (Chinese proverb). The crux of a successful PBL approach is, of course, finding meaningful and interesting problems. So what makes a good problem? Here is where we will teach you, the reader, to fish for good problems as well as be able to write your own. As mentioned in Section 1, good problems should have the following characteristics:

- They take several steps to solve.
- More than one approach can be used to arrive at a complete solution.
- Good problems lend themselves well to collaboration with peers.
- Meaningful problem solving promotes flexibility of thought and innovation.
- Mathematical learning and reasoning are integral to the process of problem solving.
- Problem solving is about working around obstacles to understand the unknown.

The ideal would be to have all six characteristics, but as we've stated before, ideals are not always reached. Search for problems that have at least two or three of the characteristics. Then, could you reword or rework the problem so it captures more of the list? Just like when we asked the question of students, *"How much time do you spend consuming things other people have made, and how much time do you spend creating something original?,"* we could ask the same of ourselves. When you immerse yourself in the creative process of either rewriting problems to make them more meaningful or creating problems of your own, it opens another door for how to think about mathematics. This will entail a lot of consumption of math problems to gain the inspiration to create your own, but the best way to get started is to take that first step! And with all first steps, having a supportive community can make all the difference. Again, start small, but try and identify one or two colleagues with whom you can meet with on a regular basis to share ideas and explore different concepts. This cross-pollination of thought and being able to communicate with others is how you create the fertile ground from where interesting problems can grow. Further, having the accountability to meet with peers and discuss interesting problems, either ones you've discovered or ones you want to create, ensures that the process will be a priority in your life and make you more open to new ideas as they come along.

An example of a problem that works well with collaborators, promotes innovation, requires mathematical reasoning, and has an obstacle to work around is the following:

Without using a calculator, find positive integers x, y, z such that $29x + 30y + 31z = 366$ (note, you are not asked to provide *all* solutions, only one triple x, y, and z to satisfy the equation).

You may think this is a difficult problem, but all it takes is the ability to *notice*. What does the number 366 make you think of and why do the numbers 29, 30, and 31 seem familiar? Hopefully, someone on the collaboration team will make the connection that there are 366 days in a leap year, and once that connection is made, then you will think in terms of how many months in a year have 29, 30, and 31 days, respectively. So one possible solution is:

$x = 1$
$y = 4$
$z = 7$

SHOW RELENTLESS CURIOSITY

Good problems start with good questions, and that means being relentlessly curious. So many times in life, we go through the motions and follow the steps without really questioning. Why do these steps exist? What if we switch the order? Can we work backwards? Change the variables? Once again, it comes back to being playful, and when you are playful, you think about things in different ways.

There was a study in 2011 conducted by the Royal Society for Medicine by Sir Ken Robinson where participants were asked a simple question, "What can you do with a paperclip?" Adults were only able to think of a dozen or so uses while kids were able to think of hundreds of possibilities. As we get older and bound by the way things should be, we get out of practice asking questions. *Why* is the sky blue? *How* can bumblebees fly? Is there only one way to prove the Pythagorean theorem? (Answer: There are hundreds of proofs, both geometric and algebraic; check out the full unit in Chapter 14, "Pythagorean Theorem Revisited," to learn more.) Another method for out-of-the-box thinking is to add constraints to the box (back to the concept of "Resilience Is Born Through Creativity," mentioned in Section 1.3, where we describe how creativity is stoked through constraints, e.g., living on a tight budget).

For example, one constraint you can add to your problem writing is to have it involve the current year, which also adds relevance for the students. However, this constraint should be significant for the statement of the problem. It is not enough, for example, to note that the year 2019 is an odd number. The following two problems were created by Dr. Andreescu to incorporate the year as the constraint (math competitions are particularly fond of these kinds of problems):

1) This problem uses the year 2019 as the constraint: Find all triples (p, q, r) of primes such that $pqr - 18(p+q+r) = 2019$.

Solution
Because 18 and 2019 are both divisible by 3; so is pqr. Thus, one of (p, q, r) is a multiple of 3, and because it's prime, it must equal 3. Without loss of generality, assume that $r = 3$. Then $pq - 6(p+q+3) = 673$, implying
$$pq - 6(p+q) + 36 = 673 + 54.$$

It follows that $(p-6)(q-6) = 727$, and since 727 is a prime, we get $p-6 = 1$ and $q-6 = 727$ or $p-6 = 727$ and $q-6 = 1$. We obtain the solutions $(7, 733, 3)$, $(733, 7, 3)$ and cyclic permutations.

2) This problem uses the year 2020 as the constraint: What are the last two digits of 7^{2020}?

Solution
It is important to note that 2020 is divisible by 4. Further, $7^4 = 2401$ and 7^8 is 2401×2401, which ends in 01 (5 764 801). Continuing, we find that 7^{4n} ends in 01 for all integers n greater than 1. Therefore, the last two digits of 7^{2020} are 01.

Added challenge! What are the last *three* digits of 7^{2020}?

> **Solution**
> We have $7^{2020} = (7^4)^{505} = (2400+1)^{505}$.
>
> We can expand $(2400+1)^{505}$ with the binomial theorem. The first two terms are $1 + 505 \times 2400$, and every term after that is a multiple of 1000. This means that $(2400+1)^{505}$ has the same last three digits as $1 + 505 \times 2400 = 1\,212\,001$. So the last three digits of 7^{2020} are 001.
>
> In other words, modulo 1000 is congruent to $505 \times 2400 + 1 = 1\,212\,001$ (binomial theorem), so the last three digits of 7^{2020} are 001.

Another obvious constraint employed by math competitions is a time limit. For example, Olympiad-level problems should take about 1.5 hours each. The American Invitational Mathematics Exam has 15 problems that need to be solved in 3 hours, approximately 12 minutes per problem. With an increasing number of mathematics competitions being held around the globe, it becomes difficult for problem authors to create novel problems that don't appear on other exams, and the same is true for teachers writing problems – this can also be said for musicians, artists, poets, and others in the creative space. Creating something new is increasingly difficult and, again, that is okay! This section is all about rewiring your brain to think differently through the creative process so that your teaching and problem solving will improve. You're looking for inspiration as a part of your daily routine, and that will create a richer experience in and out of the classroom.

You want to constantly be in a "what if" mindset, and one way to document your journey of questioning is to keep a notebook handy where you can record your questions and thoughts. It's always better to use a hard-copy notebook; there is plenty of research that shows physically writing increases retention, but we live in busy times and everyone usually has their phone with them. If the memo or notes feature on your phone or computer is your preferred choice, then by all means, use it. Whatever works best for you and allows this creative process to become a habit is the recording method you should choose.

DON'T FORGET GRACE

When you first start writing problems, you may not always like the outcome, and you'll be even more critical of early problems you write as your skills improve. Further, it is inevitable that you will create some problems that you are unable to solve. All of this is part of the journey, so don't let it discourage you from the process.

Dr. Andreescu was presented with a gift from a colleague who had collected some of the problems that he had written when he was still in middle school and high school in Romania, and upon first glance, Dr. Andreescu was critical of the caliber of the problems. Yet, he gave himself grace and took time to note that what he wrote so long ago still possessed the "spark" he looks for when writing problems today – namely, that they have some interesting core idea and, of course, as his knowledge and skills increased, his problems became better and better.

Another part of the process is that sometimes the problems you write are just unsolvable, or you don't yet possess the mathematics background required to solve what you create, and that's okay! Give yourself the grace to understand you won't know how to do everything and make sure to enjoy the creative exercise. By recording your process in a notebook, you can always go back to the problems you have started, and as your skills grow and your insights deepen, you may be able to rework and/or solve them – it all requires practice, patience, and persistence to keep going! It also requires quite a bit of time solving problems written by others, with the understanding that not all good problem solvers are necessarily good problem authors at the beginning. Remember, just because someone can play music doesn't mean they can instantly compose music; however, if you make the time to create, you will have a richer understanding and can make larger leaps in ability.

Be sure to tap your support network: Your students, colleagues, and friends can help beta test your ideas as well as provide the support to keep going! You don't want to get bogged down in perfection by writing the best problem. A messy problem that your community can help polish up is just as valuable, if not more so.

Utilize Teacher Resources

Having the right tools for the job can make work more efficient as well as pleasant. There is a lot to absorb when embarking on a problem-based approach, and this section intends to provide you with the necessary resources, so the information is in one handy location. This kit includes:

- Classroom environment
- Knowledge bank

CLASSROOM ENVIRONMENT

Running a PBL classroom takes experience, so the habitat the students occupy needs to be designed and this section will help teachers new to this approach shorten their learning curve. The first thing to be understood is that you are not a judge of work that is either correct or incorrect, but instead, you are a facilitator on the path of discovery and truth. This means helping guide the students with new concepts as well as reinforcing the knowledge they have already procured.

To combat math phobia and "fight, flight, freeze," you want to create an atmosphere where there is no fear of mistakes, only the quest for truth. This means instead of correcting work, you are asking the student and/or the class, "can this be true?" or "is there another path that gets us closer to truth (the solution)?" I know this may seem like splitting hairs by changing the wording, but the *facilitator of truth is leading students to where you know they can be instead of telling them where to go* – the distinction is a critical one to have a successful PBL environment.

This means, that as the facilitator, you need to be knowledgeable, encouraging, and most of all, patient. It's the easier path to just show students how to solve a problem;

however, this hinders retention as well as the ability to apply the concept to other problems. When students struggle, that is when they grow and learn, and when you struggle in a supportive environment, you are more willing to be vulnerable and have false starts on your quest for the solution.

 When interviewing teaching assistants for the AwesomeMath Summer Program, one of the applicants was a two-time gold medalist for the US Team that competed at the International Mathematics Olympiad. Obviously, the applicant had the necessary math skills for the job, so the next concern was did he have the right approach when working with students? When asked, "What would you do to help a student who is stuck on a problem?" his answer was, "I would never be so cruel as to spoil the solution for them. I would ask them questions to help them work through the problem on their own." We certainly hired him, and he was an inspiration and incredible mentor for the kids, because he respected their abilities and understood the importance of facilitating learning.

It's not only important for teachers to move away from judging work as right or wrong, the students need to be encouraged to leave this fixed mindset at the door, but they also need to keep students focused on the journey and not the destination. Many times, the students will want their teacher to tell them if their work is correct, if the problem will be on a test, and/or whether or not they absolutely need to know the information. As a facilitator, instead of answering these questions, you will instead question them. Do *you* think the answer is correct, and if so, how could you prove it? How do you think that as a teacher, I would choose to test students' knowledge? When do you think you could apply this concept in the future? Teachers need to be able to ask questions and bring the class together for meaningful discussions. This may take longer, but the payoff will be so much greater.

Teachers will argue that they just don't have enough time to truly implement a PBL curriculum, but those who have already done so have reaped the benefits of the approach. In other words, a broadcast approach may be more efficient, but studies have shown (as have test scores) that the method is not effective long term. When teachers guide students on the path of learning at their own pace in an inquiry-based environment, they learn more deeply. This means the teacher needs to be knowledgeable and encouraging, while also allowing ample time for students to work (and struggle) with the concepts and problems.

The problems themselves should be challenging, interesting, and yet doable, while also illustrating key ideas of the material that needs to be covered. Having problems that are scalable for all levels and/or able to be "chunked" into more manageable pieces as well as have multiple approaches to the solution will increase class discussion and connection.

The following problem was created for the Purple Comet! Math Meet and was therefore created to be worked on by multiple people since this is a team-based competition.

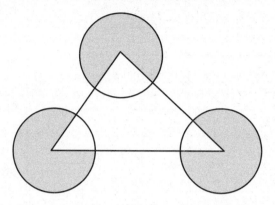 A triangle with side lengths 16, 18, and 21 has a circle with radius 6 centered at each vertex. Find n so that the total area inside the three circles but outside of the triangle is $n\pi$.

The answer is 90. Each circle has area $6^2\pi = 36\pi$. The total measure of the angles in the triangle is 180°, so the region inside the circles and inside the triangle equals the area of a semicircle with radius 6. Thus, the required area is $3 \times 36\pi - \frac{1}{2} \times 36\pi = 90\pi$. The requested value is 90.[1]

It also lends itself well to being chunked; namely, you can reason out the individual steps, or chunks, that lead you to the solution. The problem requires the student to use multiple pieces of information from their knowledge base and combine them to calculate the complement of the area, which is a concept used frequently in mathematics.

KNOWLEDGE BANK

This section is meant to provide the teacher with a framework of foundational knowledge that they will need to successfully implement a problem-based curriculum. As a framework, it will not be a comprehensive resource; rather, it will show you what areas are required so that gaps in knowledge can be discovered and filled. Take note, this will go beyond the standard school curriculum, e.g., what you would need to pass an algebra class, and instead will show the additional skills you need to implement a problem-based algebra unit.

Fractions, decimals, and percentages.

Fractions	Decimals	Percentages (%)
1/2	0.5	50
1/3	0.3333	33.3
1/4	0.25	25
1/5	0.2	20

Fractions	Decimals	Percentages (%)
1/6	0.1666	16.6
1/8	0.125	12.5
1/10	0.1	10
1/12	0.0833	8.3
1/16	0.0625	6.25
1/20	0.05	5
1/25	0.04	4
1/50	0.02	2
2/3	0.6666	66.6
2/5	0.4	40
3/4	0.75	75
3/5	0.6	60
3/8	0.375	37.5
4/5	0.8	80
5/6	0.8333	83.3
5/8	0.625	62.5
7/8	0.875	87.5

Prime numbers less than 1000.

2	3	5	7	11	13	17	19	23	29	31	37
41	43	47	53	59	61	67	71	73	79	83	89
97	101	103	107	109	113	127	131	137	139	149	151
157	163	167	173	179	181	191	193	197	199	211	223
227	229	233	239	241	251	257	263	269	271	277	281
283	293	307	311	313	317	331	337	347	349	353	359
367	373	379	383	389	397	401	409	419	421	431	433
439	443	449	457	461	463	467	479	487	491	499	503
509	521	523	541	547	557	563	569	571	577	587	593

599	601	607	613	617	619	631	641	643	647	653	659
661	673	677	683	691	701	709	719	727	733	739	743
751	757	761	769	773	787	797	809	811	821	823	827
829	839	853	857	859	863	877	881	883	887	907	911
919	929	937	941	947	953	967	971	977	983	991	997

Squares of integers from 1 to 60.

n	n^2	n	n^2	n	n^2
1	1	21	441	41	1681
2	4	22	484	42	1764
3	9	23	529	43	1849
4	16	24	576	44	1936
5	25	25	625	45	2025
6	36	26	676	46	2116
7	49	27	729	47	2209
8	64	28	784	48	2304
9	81	29	841	49	2401
10	100	30	900	50	2500
11	121	31	961	51	2601
12	144	32	1024	52	2704
13	169	33	1089	53	2809
14	196	34	1156	54	2916
15	225	35	1225	55	3025
16	256	36	1296	56	3136
17	289	37	1369	57	3249
18	324	38	1444	58	3364
19	361	39	1521	59	3481
20	400	40	1600	60	3600

Cubes of integers from 1 to 30.

n	n^2	n	n^2	n	n^2
1	1	11	1331	21	9261
2	8	12	1728	22	10468
3	27	13	2197	23	12167
4	64	14	2744	24	13824
5	125	15	3375	25	15625
6	216	16	4094	26	17576
7	343	17	4913	27	19683
8	512	16	5832	28	21952
9	729	19	6859	29	24389
10	1000	20	8000	30	27000

The first 40 Fibonacci numbers.

n	F_n	n	F_n
0	0	20	6765
1	1	21	10946
2	1	22	17711
3	2	23	28657
4	3	24	46368
5	5	25	75025
6	8	26	121393
7	13	27	196418
8	21	28	317811
9	34	29	514229
10	55	30	832040
11	89	31	1346269
12	144	32	2178309
13	233	33	3524578
14	377	34	5702887

n	F_n	n	F_n
15	610	35	9 227 465
16	987	36	14 930 352
17	1 597	37	24 157 817
18	2 584	38	39 088 169
19	4 181	39	63 245 986

The first 21 factorials.

n	$n!$
0	1
1	1
2	2
3	6
4	24
5	120
6	720
7	5 040
8	40 320
9	362 880
10	3 628 800
11	39 916 800
12	479 001 600
13	6 227 020 800
14	87 178 291 200
15	1 307 674 368 000
16	20 922 789 888 000
17	355 687 428 096 000
18	6 402 373 705 728 000
19	121 645 100 408 832 000
20	2 432 902 008 176 640 000

Pascal triangle.

```
                              1
                           1     1
                        1     2     1
                     1     3     3     1
                  1     4     6     4     1
               1     5    10    10     5     1
            1     6    15    20    15     6     1
         1     7    21    35    35    21     7     1
      1     8    28    56    70    56    28     8     1
   1     9    36    84   126   126    84    36     9     1
1    10    45   120   210   251   210   120    45    10     1
```

Rules of Divisibility

When a number is divisible by:

- 2 – the last digit is even
- 3 – the sum of the digits is divisible by 3
- 4 – the last two digits from a multiple of 4
- 5 – the last digit is 5 or a 0
- 6 – must be divisible by 2 and 3
- 7 – three-digit "zig-zag" sum
- 8 – the last three digits from a multiple of 8
- 9 – the sum of the digits is divisible by 9
- 10 – the last digit is 0
- 11 – the "zig-zag" sum must be divisible by 11; ex. $927\,498 = 9 - 2 + 7 - 4 + 9 - 8 = 11$ is divisible by 11
- 12 – must be divisible by 3 and 4

Pigeonhole Principle

If $kn + 1$ objects are distributed among n boxes, one of the boxes will contain at least $k + 1$ objects.

Count multiple events. When counting multiple events, we multiply the number of ways that A and B can occur in the following way: (Number of ways A can occur) × (Number of ways B can occur).

Find permutations. We can order n different objects in $n!$ ways where

$$n! = n \times (n-1) \times (n-2) \times \ldots \times 3 \times 2 \times 1$$

We can select r items in order from a group of n items in $P(n,r)$ ways, where

$$P(n,r) = n \times (n-1) \times \ldots \times (n-r+2) \times (n-r+1) = n!/(n-1)!$$

Basic Algebraic Identities

For all real numbers $a, b, c, d, x, y, z, t, a_1, a_2, \ldots, b_1, b_2, \ldots$

1. $a^2 - b^2 = (a-b)(a+b); \; 4ab = (a+b)^2 - (a-b)^2$
2. $(a^2 + b^2)(x^2 + y^2) = (ax - by)^2 + (bx + ay)^2$
3. $(a^2 + b^2 + c^2 + d^2)(x^2 + y^2 + z^2 + t^2) = (ax - by - cz - dt)^2 + (bx + ay - dz + ct)^2$
$\qquad + (cx + dy + az - bt)^2 + (dx - cy + bz + at)^2$
4. $a^3 \pm b^3 = (a \pm b)(a^2 \pm ab + b^2)$
5. $x^3 + y^3 + z^3 - 3xyz = (x+y+z)(x^2 + y^2 + z^2 - xy - yz - zx)$
$\qquad = \tfrac{1}{2}(x+y+z)\left[(x-y)^2 + (y-z)^2 + (z-x)^2\right]$
6. $x^3 + y^3 + z^3 = (x+y+z)^3 - 3(x+y)(y+z)(z+x)$
7. $a^4 - b^4 = (a-b)(a+b)(a^2 + b^2)$
8. $a^4 + b^4 = (a^2 + b^2 - ab\sqrt{2})(a^2 + b^2 + ab\sqrt{2})$
9. $a^5 - b^5 = (a-b)(a^4 + a^3b + a^2b^2 + ab^3 + b^4)$
10. $a^5 + b^5 = (a+b)(a^4 - a^3b + a^2b^2 - ab^3 + b^4)$
11. $(1+a)(1+a^2+a^4) = 1 + a + a^2 + a^3 + a^4 + a^5$
12. $a^5 + b^5 = (a^3 - 2ab^2)^2 + (b^3 - 2a^2b)^3$
13. $a^n - b^n = (a-b)(a^{n-1} + a^{n-2}b + \ldots + ab^{n-2} + b^{n-1})$
14. $a^{2n} - b^{2n} = (a^2 - b^2)(a^{2n-2} + a^{2n-4}b^2 + \ldots + a^2b^{2n-4} + b^{2n-2})$
15. $a^{2n+1} + b^{2n+1} = (a+b)(a^{2n} - a^{2n-1}b + \ldots - ab^{2n-1} + b^{2n})$
16. $(2^{2n+1} + 2^{n+1} + 1)(2^{2n+1} - 2^{n+1} + 1) = 2^{4n+2} + 1$

17. Lagrange's Identity:

$$\left(\sum_{i=1}^{n} a_i^2\right)\left(\sum_{i=1}^{n} b_i^2\right) - \left(\sum_{i=1}^{n} a_i b_i\right)^2 = \sum_{i,j=1}^{n} (a_i b_j - a_j b_i)^2$$

18. Catalan's Identity:

$$1 - \frac{1}{2} + \frac{1}{3} - \frac{1}{4} + \ldots + \frac{1}{2n-1} - \frac{1}{2n} = \frac{1}{n+1} + \frac{1}{n+2} + \ldots + \frac{1}{n+n}$$

Rules for Powers with Nonnegative Integer Exponents

1. $(+a)^n = +a^n$
2. $(-a)^{2n} = +a^{2n}$
3. $(-a)^{2n+1}) = -a^{2n+1}$
4. $a^m \times a^n = a^{m+n}$
5. $a^m : a^n = a^{m-n}$, $a \neq 0$
6. $a^m \times b^m = (a \times b)^m$
7. $a^m : b^m = (a/b)^m$, $b \neq 0$
8. $1/a^m = (1/a)^m = a^{-m}$, $a \neq 0$
9. $(a^m)^n = a^{mn} = (a^n)^m$
10. $a^0 = 1$, $a \neq 0$
11. $0^n = 0$, $n \neq 0$

Important Averages of Positive Real Numbers

Arithmetic mean (AM): $\dfrac{a_1 + a_2 + \ldots + a_n}{n}$

Geometric mean (GM): $\sqrt[n]{a_1 a_2 \ldots a_n}$

Harmonic mean (HM): $\dfrac{n}{\dfrac{1}{a_1} + \dfrac{1}{a_2} + \ldots + \dfrac{1}{a_n}}$

Quadratic mean (QM): $\sqrt{\dfrac{a_1^2 + a_2^2 + \ldots + a_n^2}{n}}$

Basic Algebraic Inequalities

1. $a^2 \geq 0$ for all real numbers a
2. $HM \leq GM \leq AM \leq QM$
3. Cauchy-Schwarz Inequality:

$$\left(a_1^2 + a_2^2 + \ldots + a_n^2\right)\left(b_1^2 + b_2^2 + \ldots + b_n^2\right) \geq \left(a_1 b_1 + a_2 b_2 + \ldots + a_n b_n\right)^2$$

with equality if and only if

$$\frac{a_1}{b_1} = \frac{a_2}{b_2} = \ldots = \frac{a_n}{b_n}$$

4. (T2's Lemma):

$$\frac{a_1^2}{b_1} + \frac{a_2^2}{b_2} + \ldots + \frac{a_n^2}{b_n} \geq \frac{\left(a_1 + a_2 + \ldots + a_n\right)^2}{b_1 + b_2 + \ldots + b_n}$$

Basic Geometry Facts

Parallelograms: A parallelogram is a quadrilateral whose pairs of opposite sides are parallel.

Properties and characterizations of parallelograms

1. A parallelogram is divided into two equal triangles by each of its diagonals.
2. The pairs of opposite sides of a parallelogram are equal.
3. The pairs of opposite angles of a parallelogram are equal.
4. The diagonals of a parallelogram intersect at their respective midpoints.
5. If the pairs of opposite sides of a quadrilateral are equal, then it is a parallelogram.

Rectangles: A rectangle is a parallelogram with a right angle.

Properties and characterization of rectangles: The diagonals of a rectangle are equal.

Rhombi: A rhombus is a quadrilateral that has all four sides equal.

Properties and characterizations of a rhombus

1. The diagonals of a rhombus are perpendicular.
2. The diagonals of a rhombus are angle bisectors of the respective angles of the rhombus.
3. If the diagonals of a parallelogram are perpendicular, then the parallelogram is a rhombus.

Squares: A square is a rectangle that has all four sides equal.

Trapezoids: A trapezoid is a quadrilateral that has one pair of opposite sides parallel. These are called bases of the trapezoid. The other two sides are called legs of the trapezoid. Notice that, according to this definition, any parallelogram is a trapezoid as well. There are books in which trapezoids are allowed to have one pair of opposite sides parallel.

Isosceles trapezoids: These are trapezoids with equal legs.

Properties and characterizations of isosceles trapezoids

1. The angles at a base of an isosceles trapezoid are equal.
2. The diagonals of an isosceles trapezoid are equal.
3. If the angles at a base of a trapezoid are equal, then it is isosceles.
4. If the diagonals of a trapezoid are equal, then it is isosceles.

Theorem of Thales: If two parallel lines intersect the legs of an angle so that the resulting segments on one leg are equal, then the segments on the other leg are equal as well.

The midline of a triangle: The segment connecting the midpoints of two sides of a triangle is called the midline of the triangle.

The midline theorem for triangles: The midline of a triangle is parallel to the third side of the triangle, and has half of its length.

Properties of the midpoints of the sides of a quadrilateral: The midpoints of the sides of a quadrilateral are vertices of a parallelogram.

The midline of a trapezoid: The midline of a trapezoid is the segment connecting the midpoints of its legs.

The midline theorem for trapezoids: The midline of a trapezoid is parallel to the bases of the trapezoid and has length the arithmetic mean of the lengths of the bases. The segment connecting the midpoints of the diagonals of a trapezoid is parallel to and has length half of the difference of the bases.

A characterization of right triangles: If a median of a triangle is equal to half of the side it connects to a vertex, then the triangle is right.

The fundamental theorem of proportional segments: The corresponding segments on the legs of an angle cut out by parallel lines are in the same proportion.

Similarity for triangles: Two triangles are called similar if the corresponding angles are equal and the corresponding segments are in the same proportion. The ratio of two corresponding segments is called the coefficient of similarity of the triangle.

Criteria for similar triangles

1. If two triangles have two corresponding pairs of angles with the same measure, then they are similar.

2. If the ratio of corresponding sides of two triangles does not depend on the sides chosen, then the triangles are similar.

3. If two sides are taken in a triangle, that are proportional to two corresponding sides in another triangle, and the angles included between these sides have the same measure, then the triangles are similar.

A remarkable property of trapezoids: If the trapezoid is not a parallelogram, then the point of intersection of the diagonals, the midpoints of the bases, and the point of intersection of the legs are collinear. If the trapezoid is a parallelogram, the first three points belong to a line parallel to the legs of the trapezoid.

A property of angle bisectors of a triangle: Every angle bisector divides the side it intersects in two segments proportional to the other two sides of the triangle.

Property of the area of a triangle: The product of the length of a side by its corresponding height is independent of the choice of the side.

Area Formulas

Parallelograms

1. The area of a parallelogram is equal to the product of a side by the corresponding height.

2. The area of a parallelogram is equal to the product of two adjacent sides and the sine of the included angle.

3. The area of a rectangle is equal to the product of two of its adjacent sides.

Trapezoid: The area is equal to the product of the height by the half sum of the bases.

Provide an Active Learning Environment

With problem-based learning, a lot can be going on as students interact with each other, present problems on the board, ask for help from the teacher, and continuously question. It can be tough, but an active learning environment is, well, active! That means trying to bring order to the chaos so that time isn't wasted and deeper learning can happen. There are a number of classroom management techniques that you can use to get the most out of the lesson and the students.

First, let's start with a personal question:

What Teacher Made the Biggest Impact in Your Life, and Why?

A common thread when people are asked this question is that the teacher believed in them, so they were challenged to be better and felt capable of reaching higher expectations. As we noted in Section 1, it's all about the journey with your student by being their guide and mentor so that after crossing the threshold to learning, they are changed deeply by the experience. It means viewing your students each as a collection of strengths and not a collection of weaknesses. It also means that the students need to be there for each other on the journey as well.

To manage the class well, the teacher should not assume a central part in the process. As Dr. Andreescu states when he is working with teachers in his graduate level education courses, "Mathematics is not a spectator sport, it is a team game," and that's why having students work in groups will make the learning process much more effective.

The groups don't compete against each other, but at the conclusion of their work, the whole class should be brought together for a thorough discussion. It's this collaboration and discussion where growth happens and the students are given agency (ownership) over their learning process. This doesn't mean that explanations are not necessary at some point, but they should be employed to *connect* disparate ideas and bring them together.

The beauty of this style is that it also accommodates nontraditional learners. Students find their role in the group and can add their ideas to the mix so that synergies happen. This is done with scalable, relatable, and interesting problems!

What Is Your Personal Approach to Teaching Problem Solving? How Do You Ensure the Kids Are Learning and the Process Is Effective?

Dr. Emily Herzig: Encouraging conversations between students during class is a cornerstone of my approach to teaching. I believe that opportunities for students to articulate ideas with their peers are vital to each student developing a personal understanding of the material. Conversations can take several forms—from assigning scaffolded problem sets to be solved in groups, to giving students 30 seconds to briefly explain an idea or definition to their nearest neighbor – and are easily adjusted to fit different classes and classroom environments. I also try to tailor the examples and applications presented to fit students' interests, which increases student engagement.

One advantage of fostering student conversations is that by making the students more comfortable talking to each other in class, they are often more willing to talk to me as well. I've found that students who would never speak up in front of the whole class to ask a question may be more comfortable asking the question of their peers or asking me as I walk the room during group conversations. It is of course incredibly useful to be able to gauge student learning on the spot by hearing from them what they did and did not understand. I also like to ask reflection questions on exams. Reading a few sentences about how students interpret the material provides valuable insight into their understanding and misconceptions, and (along with the in-class conversations) reinforces the importance of communication of ideas in math.

So what are the main techniques required for classroom management that mitigates chaos and maximizes collaborative learning? The list that follows are some tried-and-true techniques, but that doesn't mean you can't innovate your own approach based on the students in your class and/or experience. Trust yourself to evaluate, reflect, pivot, and innovate.

Eight classroom management techniques for active learning.

Technique	Example
Model the behavior. Actions speak louder than words, so be the curious, connected, and collaborative learner you'd like your students to be.	The figure below was made by gluing together 5 non-overlapping congruent squares. The figure has area 45. Find the perimeter of the figure. This is a fun problem that is collaborative and sparks curiosity! The figure was made by gluing together five nonoverlapping congruent squares. The figure has an area of 45. Find the perimeter of the figure. **Solution:** The total area is 45, so the area of each square is 9, and therefore the side length of a square is 3. The border of the figure is composed of 11 segments of length 3, and 2 unknown segments whose total length is $6-3=3$. Thus, the perimeter is $11 \times 3 + 3 = 36$.[2]

Technique	Example
Meditate. When you notice a loss of focus, taking just one minute to breathe rhythmically with the class can offer a much needed reset.	Have your students close their eyes and start doubling numbers in their head *slowly* as a way to create rhythmic breathing (each breath should take five to six seconds). Breathe in 1, breathe out 2, breathe in 4, breathe out 8, breathe in 16, breathe out 32, breathe in 64, etc.
Be proactive and capture attention. It's much more effective to proactively stop potential disruption than to react when it actually happens. Having students develop a mission statement and core values list will help set the proper mindset right from the start that kind collaboration is the goal for every class. When they have something fun and quick to work on from the very start as they enter the room, transitions will be smoother.	$\begin{array}{\|c\|c\|c\|} \hline 6 & 1 & 8 \\ \hline 7 & 5 & 3 \\ \hline 2 & 9 & 4 \\ \hline \end{array}$ → 15, 15, 15, 15, 15, 15, 15, 15 Capture your students' brains from the beginning with thought exercises that promote kindness, curiosity, and community. An example would be to introduce the concept of magic squares and have each student make their own and share with the class. Every row, column, and diagonal will have the same sum. Here is a simple example that sums to 15 with all the consecutive numbers 1–9. (Be sure to check out the mini-unit on magic squares to delve deeper into this topic!)
Be authentic, honest, and positive. Solving interesting math problems is a quest for truth, so be positive, honest, and authentic with your students in all your interactions. That means letting them see you struggle and question in a positive way.	Games are a great way to level the playing field and create positive and authentic interactions to promote thinking. There are a number of games in Chapter 7, but one that is quick, easy, and conducive to active learning is Buzz. Players take turns counting up from 1: the first player says "One," the second player says "Two," and so on. If a player is about to say a multiple of seven (like "fourteen") or a number with seven as one of its digits (like "seventeen"), they must say "Buzz!" instead of the numbers. Keep counting as high as you can without making a mistake or slowing down. For example, 1, 2, 3, 4, 5, 6, Buzz, 8, 9, 10, 11, 12, 13, Buzz. . . If someone doesn't "buzz," you must start back from 1.

CHAPTER 6: FIVE STEPS TO PROBLEM-BASED LEARNING 95

Technique	Example
Praise. Encourage students when they show initiative, innovate, and leave their comfort zones. When things are going well, we sometimes forget to stop and be grateful, so make it part of the routine to look for the positive and take note of it.	Consider having a chart in the classroom to keep track of students who go above and beyond by solving a problem on the board, helping a classmate, writing a problem, etc. Take the time to notice the good things whenever they happen.
Team up. Create peer-tutoring alliances so stronger students can help classmates who need more foundational help. Having an "accountabili-buddy" will let students know they have support, and stronger students will learn information more deeply in these tutoring roles.	The diagram below shows a large square with each of its sides divided into four equal segments. The shaded square whose sides are diagonals drawn to these division points has area 13. Find the area of the large square. 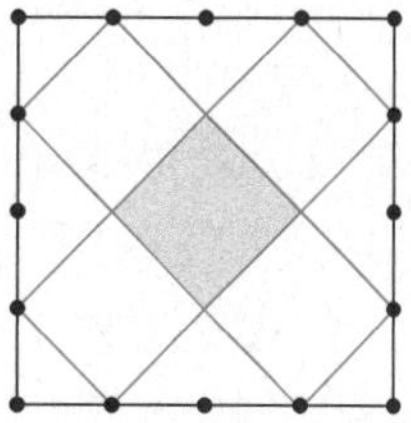 This is a fun problem that a peer tutoring team can work on together. There are two solutions provided that highlight different skills sets. The diagram shows a large square with each of its sides divided into four equal segments. The shaded square whose sides are diagonals drawn to these division points has area 13. Find the area of the large square. **Solution 1:** Let x be the area of the tiny right triangle in each corner. The shaded square has area $4x$ and the large square has area $32x$. Thus, the area of the large square is $13 \times 32/4 = 104$. **Solution 2:** By tessellating the large square, we get a diagonal grid of squares in which 1 out of 8 is shaded. Thus, 1/8 of the large square is shaded, so its area is $13 \times 8 = 104$.[3]

Technique	Example
Rotate. Make sure students rotate their collaboration teams so they can experience different approaches and avoid forming cliques. When students are working on collaboration teams, ensure that the student leader for each group also changes. When you are in a leadership role, it means being more connected with the process because you are now responsible for the team's effectiveness as well as your own. Beyond the leadership role, there are other roles students can take ownership of to ensure the group is effective. Many times, the roles are determined by the personal characteristics of each individual, but if possible, students should strive to wear different hats so that they can grow in different environments with varied expectations. Roles can include idea facilitator, innovator, questioner, thinker, and more.	Be sure to review the *Qualities of a Good Leader* later in this section. You can run a leadership fire drill game, where the leader is the one holding an object and whenever you ring a bell, they have to pass the object to another table mate who then takes over the role with the goal of solving as many problems as they can during the class problem solving session. This works best when the problems are of various difficulty levels.
Have FUN! Create an environment where kids enjoy the process along with you.	Active learning should be engaging, thought provoking, and collaborative. If you aren't enjoying the process, your students won't enjoy the process.

When coaching MATHCOUNTS teams, it was important to have a division of labor since time is of the essence in solving team-based problems. The role of team captain was pretty straightforward, and our group discussed and agreed ahead of time on the individual best suited for this role. Then, there are less-defined roles that come about more organically. For example, one team member had a particular problem-solving skill set, namely, he excelled at number theory problems, so problems of this type were directed to him so as to work toward his strengths. Further, there is always the need for a team member who keeps the team connected and acts as an arbiter of sorts to make sure the team is happy and cohesive. Sometimes this role is filled by the team captain, but in our case, it was more effective to have the role filled by another team member.

Understand the Value of Mistakes

There are wonderful ways to guide and facilitate learning that don't judge or evaluate, because you *want* your students to make occasional mistakes. If they are fearful of being wrong, it completely defeats the purpose. Mistakes are portals to knowledge and shed light on new ways of thinking and approaches.

The word *mistake* conjures negative responses from the get-go and makes a student feel as if they are personally being judged, even if you are only evaluating their homework. This is even more apparent in math courses. Humanities classes have utilized peer review and constructive criticism in education for years without the same level of negative connotations and fear of appearing "stupid." Instead, mathematics needs to be considered a quest for truth so that the results of a student's efforts can be separated from them personally, so they don't feel their intellect is being judged. In lieu of pointing out mistakes, it is more beneficial to ask questions about the truth of their responses:

- Why do you think this solution works?
- Is there another way to think about this problem?
- How would you describe this solution to your friend?
- Does this approach work for this simpler example? Why or why not?
- How would you connect your problem-solving approach with the lesson?

More lesson-specific questions can be asked as well, and in Chapter 5, we provided prompts to help you when working with students.

As you go through these questions with a student and when you recognize that a student has indeed made an error, especially ones involving reasoning, not a mere arithmetic mistake, it's important to not interrupt the student right away and instead give them the opportunity to correct themselves. It's about gently guiding and not instantly pointing out when they are on the wrong path (judging). With patience and the proper questions, they can figure out for themselves if their solution is true.

Through this quest for truth, you are training students to *notice* and self-reflect:

- Am I on the right path to solve this problem?
- Will this solution work?
- Does the solution tie back to the lesson?
- Is my strategy sound?
- Can I describe the solution clearly?
- Will I be able to present the solution to the class?

PRESENTING PROBLEMS ON THE BOARD

Much of PBL revolves around *students as teachers*, meaning that they need to be presenting problems on the board as a regular part of every class. This is an incredibly difficult thing to do, especially for some students, and even for some teachers! Presenting, processing information, and doing arithmetic in front of an audience of peers is nerve-racking. That is why setting the tone for a kind and helpful environment is so critical.

There is more going on than just solving the problem on the board. Students are learning critical life-long skills:

- **Effective communications**. This is required in business, academia, and life. Can you share your ideas with others and, better yet, the story along with your ideas?
- **Confidence**. The more practice students have explaining their work in a supportive environment, the more they can carry that skill into the future with the confidence that they know how to succeed.
- **Grace under pressure.** Public speaking is in the top three of biggest fears, so learning the coping mechanisms to have grace under pressure with a small classroom audience is essential.

As a student presents, it is tempting to help by correcting mistakes *as they happen*, but patience is still required even when they are in the more stressful situation of publicly solving a problem. Again, it's about reaching truth and not trying to save the student from mistakes – you need to put the emphasis on the process and not the student. The teacher should let the student persist in the mistake for a while until one of the classmates or the student themselves remark on it. If the mistake is kindly corrected by the group, that is preferable to the teacher as an authority figure commenting. The class is a team, and working through errors together offers a great teaching moment where everyone will benefit.

The struggle of working through a problem is so important, and yet you don't want your student to feel embarrassed or humiliated. That's why letting them know their class is there to provide help and guidance means they can focus on the problem itself and not whether they are worthy to solve it. They *are* worthy by the mere fact of being up there doing the hard work and having a support system, you and their classmates, who want them to succeed.

Recognize That *Everyone* Is Good at Math

In this book, so much emphasis is placed on process over outcomes – and that is even more true with mathematics education as a whole. There is no such thing as being "good at math," as if it's an inborn trait. It's a function of effort, and while there are some people who seem to have a natural talent, that is only because they feel passionate about the topic and therefore put in more time. It's similar to someone being a natural artist. What people don't see is the countless hours spent drawing, sketching, and painting that shapes the artist's talent so that they can produce beautiful compositions.

NUTURE TALENT

As a mathematics teacher, you are a nurturer of talent. When a student is struggling and exclaims, "I'm not good at math!" you can explain that output is a function of effort: "If you put in the time, I know you can do this!" and provide hints, encouragement, and connection (relatedness). You'll be combating a culture where it is okay to give up on math; therefore, the connection to real-life problems is key. The student needs to understand that this is for their future.

It can be demotivating as a teacher to constantly sell your product, i.e., a strong mathematics education. It may mean research on your part to ensure you believe in what you are selling. Curiosity, vulnerability, courage, and practice will ensure that you are on the right path to offer the most to your students and reap the benefits of a collaborative and supportive class of kids.

RELATE, REFLECT, REVISE

Teachers *and* students need to go through this exercise: How does the math relate to other areas, reflect and gather feedback on how well the lesson worked, revise and iterate to improve, and then share what you've learned with your class and/or community. It's a constant process – relate, reflect, revise – and will soon just be a part of your everyday teaching.

The Cornell method[4] of note-taking utilizes this technique and is strongly recommended for the teacher and for the students. It's a specific process, but effective when you want to relate knowledge to other areas, as well as be able to apply what is learned in various ways. The Cornell note-taking system was created by Walter Pauk, an education professor at Cornell University, as a framework for students to organize, synthesize, and better learn from their notes.

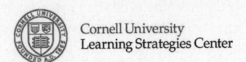

420 CCC
Garden Ave Extension
Ithaca, New York 14853-4203
t. 607.255.6310
f. 607.255.1562
www.lsc.cornell.edu

The Cornell Note-taking System

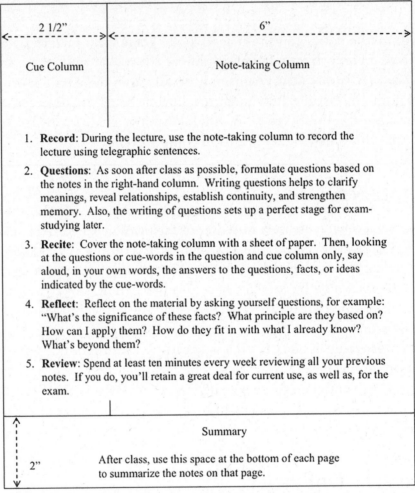

Adapted from How to Study in College 7/e by Walter Pauk, 2001 Houghton Mifflin Company

In the figure, students have a "cue column" and a "note-taking column" to *record* information. In a mathematics class, the cues can be important concepts, formulas, and ideas, e.g., they can write down the set theory notation symbols and what they mean for the mini-unit in Chapter 8 on "Number of Elements in a Finite Set" such as $A \cap B$ = intersection or $|A|$ = cardinality. The note-taking column is wider because this is not only for lecture notes, but also a space to relate, reflect, and revise. *Encourage students to draw figures and pictures when they record information as it will tap different areas of the brain to create a fuller picture of what is being learned.* This is how you synthesize complex mathematics concepts during the lecture, either a recorded lecture if using a flipped classroom approach or during class, so that the information is retained based on an individual student's learning style.

In the second bullet point in the example, *Questions* is where you take the time to make connections and relate what is learned to what you already know as well as make connections to other areas. It's an opportunity to *think* about what is being learned and *notice* patterns, discoveries, and strategies. Students organize their thoughts as well as the information and make necessary connections that lead to deeper understanding. It's important that they are prepared ahead of time for what is being covered in the class either through watching/reading the lecture notes for the class the day before (flipped classroom) or having a study sheet provided ahead of time. It is much easier to synthesize knowledge if students are given more time to process the new information. There are relate/reflect/revise questions for teachers and students at the beginning of the mini-units in Chapter 8 to help guide you through this process.

And remember, mathematics is best learned actively, not passively. This method requires interaction and thought to be successful. The *Recite* bullet point, the third step, is where students check their recall of the information in the cue column. Do they understand how and why the formulas work? Can they explain the process to a peer? Could they solve the problem at hand on their own? *There is a huge difference between understanding how something works and deriving a solution on your own.* At the AwesomeMath Summer Program, when students reach various skill levels, they delight in solving Lemmas on their own (subsidiary or intermediate theorems used in an argument or proof) that already have published proofs to test their abilities and see if they can do it. The *Recite* step is a great place to try this type of challenge.

To retain knowledge for longer and truly grasp foundational material, step 4, *Reflect*, is when you attach what is learned to a greater fabric of knowledge giving it meaning and purpose. Mathematics prowess builds step-by-step over time, and if you take the time to reflect on each of these steps forward, you will connect more with the material and be able to question its overall relevance in your life.

Our brains can only hold on to so much new information at a time, and with the pacing requirements of various districts and schools, taking some time once a week to *Review* ensures that the students don't just "pump and dump" information – meaning they take in new concepts for short-term proficiency and then quickly dump the information for the next new lesson. This is when students can also *revise* their notes based on the other lessons given throughout the week. As they learn more, they will be able to find more connections and patterns that they will see are important to the process as a whole. The Review step can either be accomplished in class (recommended) or stressed as something that needs to be done for homework. At the middle school levels, making sure that the students practice all five steps by including them as part of the classroom time will help prepare them to follow this process independently when they are in high school or college.

Finally, the *Summary* section at the bottom of the page is a space where students can write one to three sentences on what they believe is the main goal of the lesson. What do they think the teacher wants them to take away from the lesson, and why?

Any time you can guide a student to relate, reflect, and revise, it will help them have better retention, think critically, be an independent learner, and, most importantly, grow their understanding and passion for mathematics and problem solving.

Notes

1. Purple Comet! Math Meet contest, high school 2018, problem 2. See https://purplecomet.org/?action=resource/oldcontests to download old contests.
2. Source: Purple Comet! Math Meet contest, Middle School, 2017.
3. Purple Comet! Math Meet contest, Middle School, 2018, problem 4, https://purplecomet.org/views/data/2018MSProblems.pdf.
4. Adapted from Walter Pauk, *How to Study in College* (7th ed.) (New York: Houghton Mifflin Company, 2001), http://lsc.cornell.edu/notes.html.

CHAPTER 7

The Three Cs: Competitions, Collaboration, Community

To have an effective and efficient problem-based learning (PBL) classroom means engaging in mathematics competitions in a collaborative environment and being a part of a larger community for support, both for the teacher and the student. Again, it's important to stress that the end goal of these competitions is not a winning score, which is certainly a positive byproduct, but its exposure to interesting problems in varied topic areas as well as the comradery and "coopertition" that competing can bring to a group. Further, it's a goal to work toward when a competition happens at a certain time.

In this section, you will learn about the competitions available for students of various ages, skills, and interests and can choose what works best in your classroom. Please note that Section III, the Full Units, will have lesson plans that suit all the competitions since the common thread is problem solving. Next, you will be guided how to create a collaborative environment in the classroom to get the most out of each allotted period. Further, there will be ideas for how to expand the math community in your school.

Competitions

WHY STUDY FOR MATHEMATICS COMPETITIONS?

There are a myriad of activities and subjects constantly vying for a student's attention, so why should they devote time and energy to math competitions? The quick reason is "for their academic future." Many universities are looking beyond SAT or ACT scores for applicants to science, technology, engineering, mathematics (STEM) programs. They want to know if you have taken the American Mathematics Competitions (AMC) or qualified for the American Invitational Mathematics Exam (AIME). But this is not the only reason students should pursue this level of math. There are five main reasons to enter math competitions:

1. Develop critical thinking and problem-solving skills.
2. Be exposed to problems in the area of discrete math, such as number theory or combinatorics, which have many applications in today's world.
3. Build confidence through the rigor of solving challenging and elegant problems.
4. Connect with a close community of thinkers.
5. And, most importantly, because it's fun!

Solving problems for competitions is indeed a fun way to connect and provides a greater purpose to the process. Competitions supply a short-term goal with long-term benefits, and all students can help their classmates improve. There are so many competitions from which to choose that a teacher can find the contests that best fit their class schedule and approach.

MATHEMATICS COMPETITIONS

Purple Comet! Math Meet. This competition takes place around the first of April during a flexible 10-day window and is available for middle school, high school, and mixed teams. There is no registration fee; it's been "fun and free since 2003." The contest is held online and is unique because it is team-based and international in scope. All answers are non-negative integers. The contest was started by Dr. Andreescu and Dr. Jonathan Kane to provide more opportunities for team-based problem solving. http://Purplecomet.org

Math Olympiad for Elementary and Middle School (MOEMS). There are two grade divisions for the MOEMS contests: E Division for Elementary grades 4–6 and M Division for Middle School grades 6–8. There are five contests per year that run every month from November through March and can be administered and graded by the teacher during class time. The membership fee is per team (35 or fewer students consist of a team). http://Moems.org

MATHCOUNTS. This is a middle school contest for students in 6th–8th grade. Schools register in the fall and there is a per-student registration fee for students going to the Chapter Competition. Teachers/Coaches run the **School Competition** sometime in January to decide who attends the **Chapter Competition** in February, where between 1 and 10 students will represent the school. The top students from each Chapter Competition advance to the **State Competition**, which takes place in March. The top four individual competitors from each State Competition advance to the **National Competition**, which takes place in May. There is both an individual and team category. http://Mathcounts.org

Paul Erdös International Math Challenge. Every month, eight problems are posted for each of three age groups: Group A, 3rd and 4th graders; Group B, 5th and

6th graders, Group C, 7th and 8th graders. Students have the chance to solve and show their reasoning for the problems, providing an opportunity for young problem solvers to try their proof writing skills. This is free contest. http://inside.gcschool.org/abacus

Math Kangaroo. This is an international competition for students grades 1 through 12. There is a per-student registration fee of $21 and the school would need to become a test center for the program. www.mathkangaroo.us

Mandelbrot Competition. This competition is for high school students and follows a short-answer format with five rounds taking place throughout the school year. The problems range in difficulty (there is no calculus) to entice novice to advanced problem solvers. www.mandelbrot.org

MathWorks Math Modeling (M3) Challenge. MathWorks Math Modeling (M3) Challenge is a contest for high school students in 11th and 12th grades. Students work on teams to tackle a real-world problem with time and resource constraints to see what the life is like for mathematicians working in industry. The Challenge awards $100,000. Teams must be comprised of three to five students and one teacher-coach with a maximum of two teams per high school. https://m3challenge.siam.org

High School Mathematical Contest in Modeling (HiMCM). This algorithm-based competition was designed to allow teams of students to improve their problem-solving and writing skills through mathematical modeling. Teams of up to four students work on a real-world problem during the 11-day contest period, then submit their Solution Papers to the Consortium for Mathematics and its Applications (COMAP) for centralized judging. http://www.comap.com/highschool/contests/himcm/index.html

American Mathematics Competitions (AMC). There are three different levels of AMC tests: AMC 8, AMC 10, and AMC 12. There is a registration fee associated with these exams, and as well as a per bundle (10 tests) fee. These exams are multiple choice, and high scorers at the AMC 10 and AMC 12 level are invited to compete in the AIME. Students cannot be over the grade level or age level for the exams; however, younger students are eligible to take them if their math skills are up to the challenge. Many times, you may have a 10th grade student who has the precalculus skills to compete at the AMC 12 level and will choose to take both the AMC 10 test and AMC 12 test. Please note, however, that the AMC 10/12 exams are given on the same day and have questions in common, so this student would have to take the "A" test at one level and the "B" test for the other level. http://Maa.org

AMC 8, which is for students in 8th grade and below, and held in November.

AMC 10, which is for students in 10th grade and below. There is a 10A and 10B test (they have different questions) held in February so students have two opportunities to compete at the AMC 10 level.

AMC 12, which is for students in 12th grade and below. There is a 12A and 12B test (they have different questions) held in February so students have two opportunities to compete at the AMC 12 level.

- **USA Mathematical Talent Search (USAMTS).** This high school level test gives students one month to solve problems, and there are three rounds to the competition. Each problem requires proof articulation, and high scorers can qualify for AIME (see below). http://USAMTS.org

- **American Invitational Mathematics Exam (AIME).** This test is for students who excel at the AMC 10/12 level. This 15-question, 3-hour exam requires that each answer be an integer number from 0 to 999. http://Maa.org

- **USA Junior Mathematics Olympiad/USA Mathematics Olympiad (USAJMO/USAMO).** These contests are for students who have excelled at both the AMC 10/12 and AIME. Selection to the USAMO is based on the USAMO index (AMC 12 Score + 10 × AIME Score). Selection to the USAJMO is based on the USAJMO index (AMC 10 Score + 10 × AIME Score). These contests require proof articulation. http://Maa.org

- **International Mathematics Olympiad (IMO).** High school students who excel at the USAMO are well on their way to being selected as one of six students to represent their country at the IMO. http://Maa.org

- **American Regions Mathematics League (ARML).** The annual ARML competition takes place during the Friday and Saturday following Memorial Day. It is held simultaneously at four locations: Penn State University, the University of Iowa, the University of Georgia, and the University of Nevada at Las Vegas. Approximately 140+ teams will participate. A team consists of 15 students, high school age or lower. There is a team round, individual round, plus relays. Costs include a team fee and per-person fee plus transportation costs. www.arml.com

There are many more mathematics competitions at the local levels and, of course, others that are not listed. Check out your local math circle to find information about other contests.

OTHER PROBLEM-SOLVING-BASED CONTESTS

The following contests also have a problem-solving-based approach and offer further opportunities for students who enjoy critical thinking, logic, collaboration, and an engaging and fun community:

- **North American Computational Linguistics Olympiad (NACLO).** The NACLO is a contest for high-school students to solve linguistic puzzles using logic and reasoning skills. No prior knowledge of linguistics or second languages is necessary; they are problem-solving challenges where students gain exposure to diverse languages and seek patterns and examples of consistency. Students are exposed to natural-language processing in this engaging linguistic problem-solving competition. http://nacloweb.org

The International Olympiad in Informatics (IOI). The IOI is the most prestigious international computing contest at the high school level. The United States of America Computing Olympiad (USACO) finalists are selected via a national competition and invited to a rigorous academic summer training camp to further improve their skills. www.usaco.org

USA Physics Olympiad. The USAPhO is an exam for US residents by invitation. Students who score well on a qualifying exam, the F = ma competition, will move up to the next level. Approximately 400 individuals will be invited to sit the USAPhO, a competition for high school students, to represent the United States at the International Physics Olympiad (IPhO) competition. http://www.aapt.org/physicsteam

The US National Chemistry Olympiad (USNCO). This is a chemistry competition for high school students whose purpose is to inspire excellence in chemistry. The qualifying exams are the Local Chemistry Olympiad competition, then the national exam, then the top 20 students are invited to the summer training camp. http://www.acs.org/content/acs/en/education/students/highschool/olympiad.html

Conrad Challenge. This program is for students 13–18 years old where they are given the opportunity to become entrepreneurs and "apply innovation, science and technology to solve problems with global impact." The challenge values and develops skills in collaboration, creativity, critical thinking, and communication. The goal is to create commercially viable innovations that have the potential to better the lives on an individual, national, and/or global level. www.conradchallenge.org

Collaboration

To get the most out of any pursuit, including competitions, having a team of supportive peers and mentors makes all the difference. And while having collaboration in the classroom sounds like a great idea, it doesn't just happen. Individuals of all ages need to be trained to collaborate effectively, and this training will last a lifetime. Before collaboration can even begin, the groundwork of a kind space where all ideas are welcome must be cultivated and modeled. The quest for information needs to supersede ego and fear. In other words, the process once again is more important than the outcome.

Let's say that you've accomplished the first step of creating a welcoming collaborative atmosphere. Now what? How do you help your students collaborate well and effectively? As with all things, practice is the best way to shape skills.

Here are five steps to creating successful problem-solving teams:

Develop trust. This can be done by starting small with simpler lesson plans or even "welcoming" situations where students practice greeting each other in a positive way at the start of class. Set up scenarios where the students can role play what they think a kind and collaborative group would look like and build from there.

Suggestion. Randomize teams of students to work together so they can build relationships. The randomization can be as simple as "talk to the student to your right" or "all students born in May, work together," etc.

Define your role. Each student can add value in their own way. It's not always about finding the solution. They can be the leader, idea facilitator/questioner, innovator, or thinker, etc. It's not about solving the problem quickly as much as creating synergy to truly understand the depth and breadth of the material and respect each other's role. These roles will take on the characteristics of the individual, and therefore you are only giving them the name of the role and then letting each student define it themselves so they can succeed.

Suggestion. Create flash cards that have one role written on each card, e.g., leader, innovator, thinker, and idea facilitator, and have students randomly draw a flash card from the pile. They then need to form their group with one of the four different roles per group. Students who don't have a group (maybe the number of students in your class isn't divisible by four) can take on rotation roles where they go from group to group and add value.

Rotate leadership. Assign different group leaders for a collaboration team so students learn to listen to each other and experience different types of group manager approaches.

Suggestion. Review the "Qualities of a Good Leader" in the below section.

Practice receiving. Listening is just as critical as providing information so that ideas can build on each other. This means noticing and guiding how students interact, their adherence to the goals at hand, and how they build consensus in their relationships.

Suggestion. Modeling this behavior is the best approach for guiding your students. Set up a system where you call on students randomly. Always point out something positive about their response, even if it is to say, "This is my favorite wrong answer, because it is a great step towards understanding." When students see they aren't being judged, and instead, being valued for all contributions, it will help create the open communication necessary for PBL to thrive.

Create a recognition system. Praise the behaviors you want to see, such as civility, respect, listening, positive leadership, good questions, and curiosity.

Suggestion. A student-managed system where they are given different-colored sticky notes that represent different valued behaviors, such as one for listening or a good question, that they can put next to a fellow student's name on a chart is a great way to promote a positive classroom environment. They should have a limited number of award stickies that they can use per day.

QUALITIES OF A GOOD LEADER

Both the teacher and the students will be in leadership roles in a PBL classroom. Whether running the collaboration team (student) or running the class (teacher), it's important to understand the qualities of a good leader. As educators, you've probably had the opportunity to take classes on leadership or at least have had the experience required to be a good leader. Younger students won't have this advantage and may need a crash course along with the necessary tools to learn to be a good leader so that they can succeed in this role.

For example, great leaders inspire others to be their best, and that happens through high expectations infused in an environment of kindness. They need to see the strengths of the team members and hold them to task for those capabilities, while having compassion for each team member. This is all about searching for the truth in problem solving, which removes personalities from the equation and shifts the focus to solving the problem.

First, it's important to remember that we all lead in different ways based on who we are and what we want to achieve. Second, being a good leader is multidimensional. In other words, there isn't one absolute formula for success, but rather, there are various qualities that make a leader worth following.

The following list comprises traits that make up good leaders, and each item has two qualities listed. Students will have to figure out where they are in terms of these two qualities. For example, do they exude more confidence or humility? Are they driven by decisiveness or purpose? How can they bring both qualities to bear so that those whom they lead will aspire to be their better selves? Following the traits will be a suggestion for how to use the curriculum units included in the book to guide your students to be good leaders:

> **Confidence and humility.** Students can smell fear. Just like public speaking, running a collaboration team can be scary, but the rewards will outweigh any reservations, and in the long run, it will help the student immeasurably as they proceed through life. As a student leader, they need to have the confidence to take on the task, so how do you get them over fear of leading? The answer is to ensure that your students are prepared to take on the role through practicing in managed settings. Preparation and practice can counteract the dread and worry and grow confidence. The flip side of this, of course, is humility. A leader shouldn't steamroll over others' ideas, but instead have the humility necessary to see that someone else may have a better approach.
>
> **Suggestion.** The "Sequences" unit has an interesting set of problems that require thinking in many different ways. The best way to complete the problem set is to make sure multiple students have a voice.
>
> **Send and receive.** Communication is a two-way street. As a leader, you need to make sure the team is on task and understands what is required, so this means a certain level of "send" has to happen. The leader also needs to be aware that it is a team effort, not their moment of glory, and as such, the team needs to feel that they are listened to and their opinions are respected.
>
> **Suggestion.** The "Pigeonhole Principle" unit requires students to write mathematical proofs, not numerical answers. The team leader has to make sure the students' argument is sound, which means listening carefully to everyone's ideas.
>
> **Innovation and creativity.** Again, these are two qualities that seem similar, and yet one can derail the task at hand if not managed properly and the other has the ability

to bring positive change to a group as long as the scope isn't too narrow. Creativity is a free-thinking process, and the mind will follow various thought paths in order to come up with new ideas; however, this may lead a problem-solving team astray if the creativity isn't somewhat bounded with certain constraints, e.g., time limits, mathematical parameters, etc. Having a set time period for brainstorming allows for creativity to flourish without taking over the whole problem-solving session. Innovation is working with what you have to create something measurable that works, so the team will stay on task, but if the scope is too limited or the structure too rigid, innovation cannot flourish. Creativity can inspire innovation, and a good leader will help the team harness focus so that both can take place.

Suggestion. The "Dissection Time" unit features problems that require very creative solutions. Let your student leaders know that they should be on the lookout for innovative approaches.

Patience and diplomacy. Problem solving takes patience, and that means having a leader willing to let the team members work at the speed and ability with which they are capable. The team may need to form smaller groups to work through a difficult problem (this works well if there is a short period of time) or the problem may need to be "chunked" into pieces to help those who are struggling. These strategies require diplomacy, and a good leader knows how to achieve consensus so that the team works together effectively and patiently to solve the problem at hand.

Suggestion. The "Pick's Theorem" mini-unit lets students solve problems using Pick's theorem, but they can also use geometry instead. Students can learn both approaches, and those who understand Pick's theorem can help anyone who doesn't.

Decisiveness and purpose. Helping your students be decisive when required is a leadership quality that propels a group forward if done with kindness and understanding. "The buck stops here" is the burden of leadership, and if a group is stuck, having a strong leader who can give direction with decisive action can really help. That means having a well-defined purpose. The purpose can be defined by the problem at hand, a strategy that should be employed, and/or a concept that must be applied.

Suggestion. "The Number of Elements in a Finite Set" mini-unit includes many problems that *require* a lot of work and arithmetic. A good team leader can cover the entire unit without running out of time or getting sidetracked.

Integrity and principles. Much of this book focuses on creating a kind and inclusive environment that is searching for truth, not just correct solutions. Leaders must model the integrity that contributes to this positive environment. That integrity can take many forms, such as respect for the process and the players, honesty in all actions, and the quest for knowledge. Having a strong bedrock of principles, which can be decided ahead of time when creating the core values document, allows each leader to work within an integrity framework agreed upon by all participants.

Suggestion. The "Polygonal Numbers" unit goes by more quickly if students work together, but it can be difficult to divide the work appropriately. Encourage the leader to ask students to come up with a plan ahead of time as a group for tackling this unit.

Vision and perspective. Leaders must have vision, and this can be difficult for students who don't quite have the emotional maturity to control their own lives and yet are expected to have vision when running a group. Just like chunking a problem, they can be guided to realistic expectations and practice this valuable skill under the mentorship of their instructor. Having vision will come easier to some kids, and others will require more work. But what is vision? It's the ability to see how goals can be reached by providing clear communication, adhering to the values of the team, and focusing the team's efforts for efficient results. This is slightly different from perspective, which means to gain a clearer view by comparison. For example, you can feel better about a team's progress when you have the perspective to see how they are improving over the course of the class (e.g., compared to how they were last week). A good leader will have the perspective to know the well-being and growth of the groups so as to set realistic goals in order to achieve their vision.

Suggestion. The "Viviani's theorems" unit asks students to discover and derive important theorems. Each problem has a distinct purpose and vision.

Motivation and accountability. Vision requires the ability to motivate your collaboration team. As highlighted in Chapter 4, in Daniel Pink's book, *Drive*, when individuals have autonomy (control over the process), mastery (skills to succeed), and purpose (meaningful work), they will be internally motivated and a leader can help nurture these characteristics. PBL needs to have meaningful problems where students enjoy the process and want to work toward mastery both in the group and on their own. Sometimes, however, motivation needs a jump start. Accountability can help provide the impetus to get the team going. Assigning "accountabili-buddies" will help the students keep each other on track while they build the resilience to tackle the tough challenges in a rigorous problem-solving course.

Suggestion. The "Factorials!" unit includes problems that are interesting but easy to miscalculate. Students will do much better with these problems if they're paired with accountabili-buddies.

Creating effective collaboration teams, while extremely beneficial for PBL, also provides useful classroom benefits such as:

- Inclusion and diversity
- Importance of trade
- Identity capital
- Mathematical games

Dr. Alicia Prieto Langarica

How Have Your Teaching Methods Evolved Over Time, and Why?

I have gone from mostly active lecture to mostly group work and activities. This change has happened mainly due to pedagogical research that shows how active learning is so much better for students, and especially for students of color. Personally, I have found that students very much enjoy group work and learn better than just by listening. Also, it helps students take ownership of their own learning, which not only helps in my class but in their college career in general.

Why Is a Collaborative Problem-Based Approach Worthwhile?

I think as teachers we tend to forget how we learned. We did not learn because someone told us how things worked. We learned because we took things apart, we got our hands dirty, and we tried different approaches. We learned because we made a lot of mistakes. And, maybe most importantly, we learned because we talked about the concepts, while working with others, and we learned because we teach!

Collaborative work helps emulate this environment and helps students that know a bit more really secure the concepts by trying to teach others. It helps students feel more comfortable making mistakes, and by doing so, learning by experimentation. It helps fight imposter syndrome among historically underrepresented groups and seeks to create a place for students to feel safe and to enjoy playing with mathematics.

INCLUSION AND DIVERSITY

By creating small collaboration groups or teams, a differentiated learning environment is organically nurtured and inclusion will take place. Students become teachers and help their peers to connect more readily with the material. When the environment is friendly, then all students can feel a part of the team. This approach also breaks kids free from the bonds of staying still in one place. They can move more freely, and it helps the "fidgeters" be more focused when they aren't feeling physically stifled.

Further, when the focus is on each child becoming their best version, they are free to learn in the way that is best for them. This reduces the stigma that can come with fitting a one-size-fits-all model. It's particularly useful for girls who thrive in a problem-based model. When running the Metroplex Math Circle, the female participants would always sit in the back of the auditorium and work together. Collaborative teams create a "back of the room" feeling where girls can work together and not be talked over, but instead, inspired, and willing to connect.

Diversity can come in many forms: academic diversity, diversity of background, socioeconomic diversity, etc. Collaboration classrooms, when initiated correctly, will seek the strengths of each individual who can then bring his or her skills to the table. Our diversity is what makes us all better and broadens our worlds.

CHAPTER 7: THE THREE Cs: COMPETITIONS, COLLABORATION, COMMUNITY

IMPORTANCE OF TRADE

Sharing mathematical insights is a type of trade, and when students engage in the trade of ideas, everyone learns more deeply and faster than they would on their own. And when students share insights, they are of course sharing other important things as well, such as curiosity, questions, and their enthusiasm for problem solving. This is contingent on having a classroom where trade is open and welcome. The synergy that happens when students share with each other is a key component of an active learning environment.

IDENTITY CAPITAL

We all only have a finite amount of time, and there are investments that must be made of that time. As adults, our day is split in spending time in the following areas:

Academic capital. Time spent learning new things

Social capital. Time spent with family and friends

Financial capital. Time spent on our financial well-being

Physical capital. Time spent on our health

We choose how much we invest in each of these areas, but in today's world of networking, social media, and the ubiquity of personal online information, students must also be guided to invest in their *identity capital*. This is a generation that is used to every aspect of their lives being recorded and shared, and that can lead them to a false sense of security. As they reach middle and high school, they need to start honing the very important skillset of social networking, which means highlighting the aspects of who they are that they wish to share. It's not about being inauthentic; in fact, it's about being more authentic to who they truly are. Students need to fight against the temptation that the world is their personal echo chamber for their latest thoughts and feelings. It means striving for civility and respect in every and all interactions. Treating encounters with the mindset that the person they are engaging with can be a future colleague and/or potential letter of recommendation is a mindset that should be adopted. What all of this really means is students learning to be kind to themselves and to others, as opposed to the easier path of judgment or comparing.

Thinking is difficult; that's why most people judge.

–Carl Jung

Teens don't have control over many parts of their lives or the life skills to truly understand what all goes into planning things such as a school curriculum, or a job, or a household, or even just making a meal – they will sit in judgment on what is wrong or could have been done better because they have enough experience to compare, but not the depth of understanding to know the complexities. This mental exercise of comparing or judging is not a bad one, but it is incomplete without understanding. PBL hinges on truly understanding the underlying complexities and beauty of a problem over quickly reaching a solution. This means removing the potentially negative reactions of judgment that can come from lack of understanding as well as fear over the new material.

However, when students are tasked with managing their *identity capital* – e.g., if your teacher were to write a letter of recommendation for you, what would they say? What qualities do you bring to the classroom? – then they start thinking about their interactions and reactions to new things in a different way. You provide the students with more control over a situation because they can always control themselves. You aren't dictating behavior, you're encouraging them to think more long term about the networks they hope to create in the future and how those networks are started and nurtured. At the AwesomeMath Summer Program, the concept of identity capital is addressed right away at opening ceremonies. The program is there to help kids from around the globe improve their problem-solving skills, so from the outset, we ask them to think about their goals in the program and what version of themselves they would like to be. Because let's face it, we all have different versions of ourselves that we can choose to highlight at various times. If you are in an environment of negative people, it is easy to have a negative version of yourself come to the surface. That's why creating a kind and inspiring environment is paramount to running a productive PBL classroom. As mentioned in Section I, a mission statement and core values will help in this regard; however, appealing to a student's personal long-term goals of networking and connection based on an investment of time in their identity capital will aid in producing a cohesive problem-solving group.

MATHEMATICAL GAMES

For stronger bonds and cohesion with PBL, making time to play is very important, but it doesn't have to be frivolous. There are so many fun games that promote mathematical thinking, and they are a great way to solidify relationships and logical reasoning, all while having fun! Plus, there are so many free games out there, you don't need to break the bank to play them.

Below is a list of games that will get you and your class started, but it is by no means comprehensive (that would take another book!), so use the list as a springboard for exploring new games. Puzzles and simulations are individual pursuits, but students can parallel play and talk to each other while doing them. Remember, there are online versions of many of these games. A couple of great websites to explore are Cut the Knot[1] and the National Library of Virtual Manipulatives[2] for online puzzles and games.

This chart will help you assess the time commitment, number of players, type, and learning goals. You'll notice in "type" there are zero-sum and nonzero sum games. In game theory, this is the mathematical representation that one player's gain is equivalent to another player's loss, so in nonzero sum games, everyone improves. One could argue that chess is a nonzero sum game even though a "versus" type by its very nature is zero sum. This is because you are still improving the more chess games you lose, but that is a stretch of the definition.

Game	Players	Type	Time commitment	Materials	Learning goals
24		Puzzle		Optional deck of cards	Arithmetic, order of operations
Buzz	2+	Versus (non-zero-sum)	A few minutes	None	Counting, divisibility
Chess	2	Versus (zero-sum)	30–60	Chess set	Logical reasoning, game theory, strategy
Chess puzzles		Puzzle		Chess set	Logical reasoning
Chomp	2	Versus (zero-sum)	A few minutes	40 or more markers, 1 a different color	Game theory, parity, invariants
Equate	2–4 or teams	Versus (zero-sum)	30–60 min	Equate game	Arithmetic, fractions, strategy
Game of life		Simulation		Computer, optional graph paper	Counting, functions, automata
Geometric construction		Puzzle		Straight-edge and compass	Geometry, problem solving
Hive	2	Versus (zero-sum)	15–30	Hive set	Logical reasoning, strategy
Logic puzzles		Puzzle		None	Logic, reading comprehension
Mancala	2–4	Versus (zero-sum)	15–30	Mancala set	Counting, arithmetic, logical reasoning

Game	Players	Type	Time commitment	Materials	Learning goals
Mastermind	2	Puzzle	A few minutes	Optional Mastermind set	Logical reasoning
Prime Climb	2–4	10 min per player	15–30	Prime Climb game	Math visualizations, arithmetic, primes
Nim	2	Versus (zero-sum)	A few minutes	10 or more markers	Game theory, parity, invariants
Northcott's game	2	Versus (zero-sum)	A few minutes	Graph paper, 3–8 markers in each of two colors	Game theory, parity, invariants
Reversi	2	Versus (zero-sum)	30–60	Reversi/Othello set	Logical reasoning (this is a great "think ahead" game)
Set	1+	Puzzle/Versus (non-zero-sum)	5–30	"Set" cards	Classification, qualification, sets
ThinkFun puzzles		Puzzle		ThinkFun puzzles or equivalent	Logic, thinking ahead, problem solving
Towers of Hanoi		Puzzle		Stackable discs or markers of increasing size	Recursion, induction

Community

COMPETITION COMMUNITY

Finding your tribe is important for all of us. In mathematics, having a community that values meaningful problems, provides support and mentorship, and has a kind culture of sharing and growth will show students that they belong to a thriving and connected group. How can you create this community when students as well as teachers have such busy schedules? Luckily, depending on your area, you don't always have to create this community on your own, especially since problem solving is at the core of competitions, clubs, and circles.

The most prestigious mathematics competition in the world is the International Mathematical Olympiad, which serves as a first-rate example of how community inspires excellence. When you are at this level of ability, however, how do you make this community come together? The answer is the Math Olympiad Program (MOP), an intensive summer training camp that invites top-scoring students of the USAMO to train together, and from that group, six students are chosen to represent the United States for the IMO. Top-scoring USAJMO students are also invited to train at MOP. Dr. Loh, the current leader of the US IMO team and a professor at Carnegie Mellon University, decided to take this MOP invitation even further and invited teams from other countries to train with the US team. In 2015, under the leadership of Dr. Loh, the US team won first place for the first time in 21 years! The last time the US team came in first place was under the leadership of Dr. Titu Andreescu in 1994. Dr. Andreescu was also the leader of the US IMO team the year Dr. Loh competed (1999) and won a silver medal. Dr. Andreescu instilled the importance of community in his Olympiad training and continued with this approach when launching the AwesomeMath programs. Students and staff travel from around the world to learn exemplary mathematics in a community that connects inside and outside the classroom. Creating a global community has allowed US competitors to take a huge leap in both mathematics skills and comradery because they are a part of something bigger than themselves. According to an interview with the *New York Times*,[3] Dr. Loh responded to a question about IMO team members from other countries training with the US team:

> *First, bringing in the international students gives the top US students peers. They always tell you, if you're the smartest guy in the room, you're in the wrong room. So we bring in these peers, who are actually at the same level as these top six. Of course that increases the level.*

Finding peers and a math community makes a huge difference in the lives of young problem solvers. And you needn't wait until you are at the IMO level to make it happen! Engaging in mathematics competitions such as Purple Comet Math Meet, MATHCOUNTS, and the AMC allows students to find peers at their level. It's not about being the best but instead, trying to be better than you were the day before. The Purple Comet is a team-based competition, so students have a community of peers with which they can compete to solve interesting problems. These peers don't necessarily have to be kids at their own school; they can expand to finding teammates, for example, at their local library or mathematics enrichment centers. Competitions such as MATHCOUNTS and AMC allow you to increase your community as you increase your skills. For example, MATHCOUNTS has school, chapter, state, and national level competitions where students can meet other mathletes who enjoy solving interesting problems and find true peers at their ability level.

Of course, there are online forums for mathematics competitions where students can tap into a global community with shared interests. The Art of Problem Solving[4] organization hosts many of these forums where students can come together and discuss competitions, problems, and find resources.

Some schools are fortunate enough to be a part of math leagues such as the ARML, where students can send teams to participate in this engaging competition and increase their community. If you don't already belong to a math league, a great first step is to create a math club in your school. Math clubs allow students to engage in a community of thinkers through an after-school program or, in some schools, math clubs, which might be held during the school day, such as through learning lunches or open class periods.

MATH CLUB COMMUNITY

A math club doesn't need to be created for the purpose of joining a math league. Instead, it can be an additional opportunity for students to be a part of a community outside the class. You can create a math club for your school, district, or community, and the method of participation is up to you. Again, today's students are very overscheduled, so fitting in a club may be a difficult undertaking, but be creative! Some options for hosting a math club include:

- After or before school
- During a study or free period
- Over lunch
- Online with a video chat program such as Skype
- Through email groups where interesting problems can be posted and discussed
- On weekends at a library or other free location

Further, a math club can run competitions as a way to work toward a goal and expose kids to interesting problems. Teachers needn't, however, shoulder the burden of managing the club alone. Tap the community to see who can help facilitate the club and provide meaningful contributions. Individuals who enjoy empowering students to succeed in problem solving include:

- Other teachers in the community
- Business leaders, especially those in STEM fields, who are willing to donate time to help local students
- Professors, instructors, and undergraduate/graduate students from local universities and community colleges
- Mathematics enrichment centers and organizations (although these may charge fees)

- Parents, many of whom have strong mathematics backgrounds and are invested in seeing their students succeed in problem solving

When Kathy ran an enrichment school for highly and profoundly gifted students, Eudaimonia Academy, she would make inquiries at the University of Texas at Dallas and the University of Texas at Arlington for PhD students who wished to practice their dissertation defenses in front of a young audience and answer questions from the kids. Being able to see young PhD students who are passionate about their research talk about mathematics in such an engaging way connected deeply with the students. The graduate students were more relatable, and the mathematics problems they were solving showed the students in the Academy where math can take them and that the journey was worth the effort.

MATH CIRCLE COMMUNITY

Many communities, especially those with universities, have the benefit of a local math circle being held in their area. As described in Chapter 2, math circles are math programs for middle and high school students offered on a periodic basis (sometimes weekly, bimonthly, or monthly) and appeal to students looking for mathematics enrichment and topics beyond what the schools offer. They are usually held on college campuses and have a problem-solving focus. The circle is a great way to increase the size of a student's community and get them involved in sharing their own knowledge. Older students at math circles many times will be mentors for younger students. To see if there is a math circle near you, visit the National Association of Math Circles[5] website. If a math circle does not exist at your local university or college, again you can approach professors in the mathematics, statistics, computer science, and/or other STEM departments to see if they would be willing to start a circle. The National Association of Math Circles offers a free resource guide, *Circle in a Box*,[6] with tips for starting a circle and free session plans.

Math circles don't have to be held on college campuses. You can find other mathematics teachers in your area and start a circle for the kids. This will also help when developing your professional learning community (PLC).

PROFESSIONAL LEARNING COMMUNITY (PLC)

Having a community to offer you support as you create a PBL classroom has multiple benefits. While it isn't essential, never underestimate the power of your own design team or PLC. This community can help you work through problems you create or curate, share ideas regarding lesson plans, offer "been there, done that" advice, and so much more. Collaborating with like-minded peers gives teachers much-needed connection, and, as mentioned previously, when you engage in the trade of ideas, everyone improves.

When seeking other teachers who value PBL, you have a number of choices:

- **Local peers.** Having other like-minded teachers, either in your school or district, allows for face-to-face interactions and sharing of resources. These relationships can grow over time and help you divide and conquer the task of designing lesson plans.
- **Online peers.** There are many online PLCs for PBL, active learning, or competition training. Any of these three areas would provide access to like-minded colleagues. And of course, the Math Teachers Circle[7] is a great place to find your PLC as well as your local math circle if there is one in your area.
- **STEM peers.** When running a PBL classroom, you can also find peers and colleagues in other disciplines who follow a similar teaching method such as teachers of Computer Science, Physics, Statistics, etc.

MAP OF ENGAGEMENT AND PARENT INVOLVEMENT

An often-overlooked and yet robust learning community is the parents of your students. A PBL curriculum can be challenging for them, since it may be different from their own education experiences. However, there is plenty of research available that shows how this approach is more conducive to creating tomorrow's thinkers than the rote or arithmetic-centric curriculums they may remember when they were in school. Making sure that parents are on board from the beginning will give them a stake in the process, but that's not enough. Parents want a "map of engagement" for how they can help their kids succeed with their mathematics homework.

PBL is a method of teaching that is inclusive because it crosses boundaries of nationality, race, socioeconomics, ability level, and mathematics understanding. This is because PBL utilizes problems that are relevant and meaningful to the workforce our students will be entering as well as the workforce in which their parents currently operate. We don't want to train students for jobs that a computer can do better. They need to be able to think and notice shifts in data and operations so that they can add value whether they work in technology, construction, agriculture, manufacturing, the service industry, or corporations.

How can parents be provided a *map of engagement* if they don't have a background in PBL? It's the same way consultants can add value to a company when they don't work for the organization – parents can bring in a fresh perspective and help their student be curious and ask good questions. When the work is about the process and not the outcomes, then you can focus on making the process the best it can be, and that means knowing how to ask the right questions. Parents know their own kids and can question them about their approach to the problem at hand. In Chapter 4, we discussed how to create a classroom environment where questions are welcome in the segment "Mastery through Inquiry." Many of these same techniques, slightly modified for the parent, can be shared so that parents can guide their kids to asking good questions about the problems they may be stuck on. Please note that parents need to be aware of what the learning objective was for the class so they aren't completely in the dark. They may not have the training necessary, but they should be able to help their students know how to question well. Also, since they aren't responsible for providing solutions or knowing how to solve the problem, it takes away a lot of the pressure they may feel and circumvent math phobia if they have it. All that is requested of them is to help their child ask relevant questions.

- Model the behavior. If your child doesn't know where to begin, give them an example of a question you would ask.
- Encourage all questions and the process to learn. Praise your child for asking any question, even if it is a simple one, since this is the path to discovery. If you are engaged, your child will see the importance of mathematics and problem solving.
- Be vulnerable and make mistakes in front of your child. Parents need to be on the journey as well, so if you are curious and fearless of making mistakes, then your child will see that it is about problem solving and not whether or not they are good at math (they are).
- Be excited about the process. If you are interested in learning something new, your child will be as well.
- Figure out where your child is struggling with the problem. Is it the wording, critical thinking, foundational knowledge? Then, you can better help your child understand the question they need to ask.
- Share your experiences. There are plenty of problems that you may come across in your own work where you can share the process of how you solved them with your child. It will connect them to the problem solving necessary in the world and open a door for asking questions.

Aspire to Inspire: Stories from Awesome Educators

Just like collaborative problem solving can help students retain more information, think critically, and inspire them to be better, creating these learning environments can help teachers as well.

At the AwesomeMath Summer Program, instructors from across the globe fly to the United States to work with students to improve their competition math scores utilizing a PBL approach. Each instructor brings their own story and experience that shapes the way they teach their classes. Further, the longer they've been teaching, the more experience they have gained, and as a result, their teaching evolves over time. We asked a number of our AwesomeMath instructors and colleagues to choose from the following list of questions and share some of their experiences with you, the reader, because as we like to say, "When you engage in the trade of ideas, everyone improves." Note that many instructors are currently college professors, so you will have a glimpse into the environments your students will be entering, and what better way to prepare those who are going to college than to learn from the problem-based approach of college professors! See what inspired them to enter the field of mathematics and share their passion for the topic, learn their approaches to PBL, and why they value collaboration.

You will see some answers that were included in other sections of the book; however, we are reprinting them here so you have all the experiences in one place. These are the questions that were sent to the AwesomeMath team and they were instructed to answer as many as they liked:

- What were your own school experiences like in your country that contributed to your love of problem solving?
- What is your personal approach to teaching problem solving? How do you ensure the kids are learning and the process is effective?
- How have your teaching methods evolved over time, and why?
- Why is a collaborative problem-based approach worthwhile?
- What are your favorite problems, and why?

Anthony Newberry, 21 Years Teaching at Hirschi High School, Wichita Falls, TX

What Were Your Own School Experiences Like in Your Country That Contributed to Your Love of Problem Solving?

Math competitions probably were the first thing in my schooling career that brought intrigue to problem solving. MATHCOUNTS led me to problems with multiple solution paths and allowed me to encounter combinatorics and a little bit of number theory. It also led our math team together to solve problems that our very dedicated math sponsors weren't sure how to solve in an efficient manner. I remember us deriving formulas that we had never known and finding out later, of course, that they already existed, but it still gave that feeling that you had created something.

It helped that I had a precalculus teacher, who was also our competitive mathematics coach, who was a visionary. The group that I graduated with was a group who wanted to be challenged academically. Our teacher figured out that she was only challenging us on about 5% of the questions she asked and determined that she'd like to do more than that. She convinced administrators to allow her to give an 80% for correctly answering 40% of the exam, and so it opened up the other 60% of the exam to ask much more challenging questions. It took a year for anyone to achieve a perfect score in her class on a test once she made the change. So, every test had a good portion of challenging questions.

In calculus class, much of our time was spent working in groups, and the discussion of problems got very intense at times. The teacher would often ask if we wanted assistance, and we would usually answer that we wanted to solve it on our own. I was very fortunate that in a US public high school in a mid-size city, I got to discuss problems with students who eventually attended MIT, Harvard, University of Chicago, Rice University, etc.

What Is Your Personal Approach to Teaching Problem Solving? How Do You Ensure the Kids Are Learning and the Process Is Effective?

Problem solving is at the forefront of everything I am trying to accomplish. My first goal is that the student can read and understand the problems, then to teach the students how to be persistent and to learn to deal with difficulties and setbacks as well as when to ask for help and when to listen.

Many times, I give the students 5–10 minutes individually to work to formulate a solution to a problem. Then, they are allowed to work with students at their table to try to finish the solution or rewrite their solution and change the wording. They often are able to spot pitfalls in their own solutions at the table or to help each other understand something better, or to phrase things better. This is a huge skill going forward in life.

Other times, I might start them in a group and write down everything they can think of that might help them start to solve the problem. I try to ensure the kids are learning by getting quick feedback in class. I check in with students as part of the whole group as well as individually. Sometimes, I tell them they must fight with the problem for 10 minutes before I will give them assistance. I vary it up. When they write solutions, I give them feedback about what I'm looking for as they try to justify their solutions. Eventually, I have them solve problems and write up their solutions as an assessment.

How Have Your Teaching Methods Evolved Over Time, and Why?

I use more inquiry-based lessons, where students will look at data or create a table of information and try to determine a relationship or pattern. Then, they often make a conjecture, test their conjecture, and then informally or formally justify their conclusions. My time in the UTeach Master's Program at the University of Texas at Austin really helped me develop some of these ideas, both directly from the program and my development that occurred as I developed friendships with colleagues in my cohort.

Some days are still skills days, but I try to really point out what the end game of the skills are. I often tell them, maybe nobody will need to know the sin 30° 10 years from now, but if you can take information and data and analyze it to help you solve a problem, that's a skill that will translate.

Why Is a Collaborative Problem-Based Approach Worthwhile?

I think problem solving is the key reason for teaching mathematics, and I like to do that in both a collaborative way at times and individually at times. I think problem-solving ability and logic are the reasons to teach mathematics, besides the pure beauty of mathematics – all are tough sells to students, but that's what I try to sell them on, because no, they may never use complex numbers after they leave my class, but they will have to learn how to operate within a construct and use logical reasoning to make decisions and solve problems.

What Are Your Favorite Problems, and Why?

My favorite problems, especially for the classroom, are problems that have multiple solution paths. It is so hard to find this type of problem, particularly ones that are a good fit for the high school mathematics classroom. I hunt for problems a lot. The other problems that I love are problems that incorporate lots of different techniques and skills in the same problem. It really reveals who is persistent and creative.

Dr. Emily Herzig, Instructor of Mathematics at Texas Christian University

What Is Your Personal Approach to Teaching Problem Solving? How Do You Ensure the Kids Are Learning and the Process Is Effective?

Encouraging conversations between students during class is a cornerstone of my approach to teaching. I believe that opportunities for students to articulate ideas with their peers are vital to each student developing a personal understanding of the material. Conversations can take several forms – from assigning scaffolded problem sets to be solved in groups, to giving students 30 seconds to briefly explain an idea or definition to their nearest neighbor – and are easily adjusted to fit different classes and classroom environments. I also try to tailor the examples and applications presented to fit students' interests, which increases student engagement.

One advantage of fostering student conversations is that by making the students more comfortable talking to each other in class, they are often more willing to talk to me as well. I've found that students who would never speak up in front of the whole class to ask a question may be more comfortable asking the question of their peers or asking me as I walk the room during group conversations. It is of course incredibly useful to be able to gauge student learning on the spot by hearing from them what they did and did not understand. I also like to ask reflection questions on exams. Reading a few sentences about how students interpret the material provides valuable insight into their understanding and misconceptions, and (along with the in-class conversations) reinforces the importance of communication of ideas in math.

How Have Your Teaching Methods Evolved Over Time, and Why?

When I began teaching, I heavily utilized a traditional lecture format. However, I quickly saw that students were better served by getting to engage in the material themselves with feedback from myself and their colleagues, and so I began incorporating active learning techniques. Personally, I have gravitated toward utilizing group work in whatever capacity the class schedule and physical space will allow. I've found that I most enjoy working with small groups and allowing students to dictate the conversation as they work through a concept.

There was (and continues to be) a lot of trial-and-error in finding active learning methods that worked for me and my classes, and there are a multitude of active learning techniques that could work for others. The single most helpful resource was joining a community of teaching professionals focused on active learning. Having a network of educators who could share ideas, feedback, material, and support opened my eyes to the various active learning options and emboldened me to try them in my own classes. For me, that community is the MAA Project NExT, which is a fellowship program in the United States for new faculty in higher education, though similar programs and Math Teacher Circles can be found for other regions and school levels.

Why Is a Collaborative Problem-Based Approach Worthwhile?
Collaboration in the classroom has many benefits. Research has demonstrated that active learning improves performance on exams, and the effect is especially large for disadvantaged students. Currently, education is not equitably accessible to all students, with students from underserved populations and first-generation college students in particular facing additional obstacles to entering, navigating, and excelling in higher education. Thus, collaborative learning in the classroom could be key to closing the achievement gap and allowing capable but underprepared students to reach greater success in math.

Furthermore, a collaborative and problem-based approach gives younger students a more accurate impression of what higher-level math entails. Students too often carry the belief that success in math is based in rote memorization and drilling problems. While those skills are certainly useful for efficiently carrying out the basic mechanics of solving problems, it is equally important that students are able to formulate and interpret more complex problems and work with their colleagues to develop and execute problem-solving strategies. Arguably, this process is also what makes math such an enticing subject. A focus on collaborative problem solving is a great way to attract students to and prepare them for careers in math.

Dr. Alicia Prieto Langarica, Associate Professor of Mathematics, Youngstown State University

What Were Your Own School Experiences Like in Your Country That Contributed to Your Love of Problem Solving?
Growing up in Mexico, I was not very interested in mathematics as most of it seemed like a lot of memorization, such as learning the multiplication tables. However, while transitioning from elementary school to middle school (sixth to

seventh grade), I attended a regularization session before classes started in which I was given my first "Olympiad" type problem. The problem was interesting, and I knew the solution could be obtained by trying all 100 cases; however, I have always been "lazy," so I remember sitting there thinking: there has to be a better way that gets me to recess sooner.

After finding a creative solution to that problem I was hooked. I fell in love with that way of thinking: how can this be done faster, easier, better? How can I look at this in a different way? How are things connected and how can I use those connections to find the answers? Middle school and high school were filled with more problem-solving opportunities. Not only in the math club, which prepared students for the Olympiads, but in my math classes in general, there were always opportunities to think in different ways and to solve hard problems with few time constraints.

What Is Your Personal Approach to Teaching Problem Solving?

Teaching problem solving is, to me, like teaching someone to throw a spiral with a football: something students have to do by trying, not by watching. As an associate professor in Mathematics at Youngstown State University, I have a lot of very different opportunities to work with students in problem solving. In class, there are three main ways I help students practice their problem-solving skills:

1. *Group exercises.* When teaching, I try my best to lecture as little as I possibly can by briefly introducing students to main concepts and then assigning problems and exercises that aim for the students to get their "hands dirty" and discover for themselves what those concepts really mean and how they fit in the mathematics field. These exercises are mostly done in groups, with me as the facilitator making sure everyone in each group has a voice and their voice is heard. Working in groups, students are more willing to explore, make mistakes, and help each other. We then take the time to openly discuss our understanding and learn from each other.
2. *Class projects.* In many of my classes, I assign a project with the main goal of having students use the concepts learned in class to solve a real-life problem. This strategy addresses many issues, such as:
   - Having students answer the question: When will I need this?
   - Helping students bridge the class with their future jobs, or their lives in general.
   - Expose students to "real-world problem solving."
   - Working in teams and learning to collaborate.
   - Learn to write math, as well as to talk about mathematics for different types of audiences.
3. *Undergraduate research.* I am heavily involved in directing undergraduate research. On a typical semester, I work with anywhere around 10–25 students mostly

on research teams in a wide variety of projects that range from math biology to data sciences. The focus of my research with students is to give them the skills to solve problems in the real world and make them more marketable and better professionals. Most times, this means I have a very hands-off approach to research, letting students decide the direction our problem should take us, what methods we should explore and know that mistakes will be made and that is actually a great thing.

How Do You Ensure the Kids Are Learning and the Process Is Effective?
Assessment, assessment, assessment. In most of my classes, there are worksheets or quizzes almost daily (most of these are in groups). I also have more than a few exams (individual) which they take once in class and then take the same test as a take home. Lecture time is very active with constant questions and doing my best to create an environment in which students feel safe to speak up and make mistakes.

How Have Your Teaching Methods Evolved Over Time, and Why?
I have gone from mostly lecturing to mostly group work and activities. This change has happened mainly due to pedagogical research that shows how active learning is so much better for students, and especially for students of color. Personally, I have found that students very much enjoy group work and learn better than just by listening. Also, it helps students take ownership of their own learning, which not only helps in my class but in their college career in general.

Why Is a Collaborative Problem-Based Approach Worthwhile?
I think as teachers/professors we tend to forget how we learned. We did not learn because someone told us how things worked. We learned because we took things apart, got our hands dirty, and tried different approaches. We learned because we made a lot of mistakes. And, maybe most importantly, we learned because we talked about the concepts, we worked with others, and we learned because we teach!

Collaborative work helps emulate this environment and helps students that know a bit more really secure concepts by trying to teach others. It helps students feel more comfortable making mistakes and by doing so, learning by experimentation. It helps fight imposter syndrome among historically underrepresented groups and seeks to create a place for students to feel safe and to enjoy playing with mathematics.

What Are Your Favorite Problems, and Why?
Ahhh! This is such a hard question. Lately, I have been teaching a class called Quantitative Reasoning, which is a class for liberal arts and social sciences majors. In this class, I assign a group project for which I give them almost the entire semester. The project question is: "What would you change at Youngstown State University (YSU), and how?" Students present their projects at the end of the

semester to an audience of their peers and members of the YSU community who have the power to make their project happen.

The results of these projects have completely blown me away. A student group proposed installing windmills, calculated how much they would cost, when would they pay for themselves, and where to place them to maximize energy production. YSU now has two windmills. Another group worked on creating a YSU app, which we now have. This semester, a group worked on getting a commuter center at our university. They figured out the cost, did research on what other universities offer and how that has affected graduation time of their students, used (basic) mathematics to analyze cost and benefits, and put up a fantastic presentation of why this needs to happen. We will now work on making it a reality.

In the past, I enjoyed helping students figure out really hard mathematics problems and understanding the beauty of the science. I still very much enjoy doing this with my math majors. However, taking students who are terrified of mathematics and having them learn to use math as a tool to make something happen in real life is my newfound passion and is incredibly rewarding. These will be students who will no longer shy away from numbers and statistics and who will, hopefully, become informed citizens of this world.

Dr. Mirroslav Yotov, Assistant Professor of Mathematics, Florida International University

What Were Your Own School Experiences Like in Your Country That Contributed to Your Love of Problem Solving?

I was born and raised in Bulgaria. I got my education there, too. My school experiences were crucial for my choosing to become a mathematician. Looking back, I think that I was lucky to have excellent teachers, to use very well written textbooks, and to attend strong math circles. My teachers were not only teaching us their subject, they were also trying to detect our strong abilities and to encourage their further development. I do not remember anything significant related to math from my elementary schools years other than it was fun for me to do math exercises, and to be almost always first in finishing them. One episode I remember well, though! When learning the multiplication tables, I was not doing well! This prompted my grandmother, who was helping me learn the tables then, to use beans in helping me compute things faster! I am glad this episode did not make me hate math.

My encounters with interesting math, and the first recognition I got for being pretty good in understanding the theory and with doing math problems, began in my middle school years. That was where I realized for the first time how important it is to have good (educated and motivated) teachers, and to read well-written books on the subject. I fell in love with geometry, which was more challenging than algebra, which I was doing easily without too much thinking. That was the time period when my math abilities were recognized. This recognition came at first with my success in mathematical Olympiads and then the strong suggestion of my math teacher to apply after the middle school to a specialized math high school. This suggestion gave me a lot of confidence and made me want to learn more in math.

My high school years were instrumental in my developing as a mathematician. I spent four years there with the last three years under the guidance of college faculty members as my mathematics teachers. These teachers prepared students for math competitions, including the IMO. During that time, I learned mathematics from the modern standpoint, e.g., Euclidean geometry taught based on the Hilbert's axioms, trigonometry, basics of linear algebra, plane analytical geometry, and the medley of topics known as algebra, but with the relevant theory thoroughly discussed. Two courses stood out: math analysis (a beginners' version) and Galois theory. Our problem-solving sessions in class were very strong. As a matter of fact, because of the class work solving interesting problems, we didn't maintain the math circle in the high school years. But I was working a lot on my own! These three years in the school were successful for me from a competition standpoint as well, since I was selected to be on the Bulgarian IMO team in 1981.

Speaking of my high school experiences contributing to my love of math and math problems, I have to confess that my first year there was also very, very important. I will always be grateful to my math teacher for actually convincing me that I was good at math! This process started almost unnoticeably for me and for my classmates when my teacher began telling me that some of my proofs were better than the ones she knew. The grand breakthrough came after a problem was solved only by me, which caused my teacher to pull me out of my literature class and asked me to explain my solution to the students in another math class! This was a public recognition that elevated my confidence and made me respected by all my classmates. After that event, I was ready to learn EVERYTHING in math. My first-year high school teacher obviously knew how to motivate students to work and do their best!

The second similar recognition came when my algebra teacher invited me to do a project in math when I was in my third year at high school. I respected all my math teachers, but the algebra one was my favorite. And when he invited me to do this project, I again wanted to learn everything about math! So, as I said, I was lucky with my math teachers.

In my current position as math faculty at the Florida International University, I am trying to mimic the exemplary teachers of my youth and show students their strengths to help and steer them to do great things.

What Is Your Personal Approach to Teaching Problem Solving? How Do You Ensure the Kids Are Learning and the Process Is Effective?

When I teach problem solving, I need to have something interesting to share with the students. This may be an interesting and unexpected idea, or a problem with an interesting formulation, or a series of problems which reveal an interesting aspect of math objects. It is very important to give the right motivation for considering the suggested problems. That's why I strive to give the context where the problem arises and explain the significance of that problem. Years ago, I was enthusiastic about problems where the solution clever, unexpected, and beautiful. I didn't care too much about the context and the motivation. Now, I think that math is not only problem solving, but also the art of asking math problems. And for this, motivation and perspective/context are very important. That is in a sense more important than having the bright idea for answering the question; the right questions prompt interesting ideas!

For the motivating portion, I usually start with introductory and explanatory (not necessarily trivial or easy) problems. I gradually steer the students toward the interesting questions to be asked. This way, they become an active part of the discovery process. This is a longer method, but in the long run, it is better, especially when working with novices. I assess the students' problem-solving progress through the discussions we have along the way and by the success they have in doing other related problems.

How Have Your Teaching Methods Evolved Over Time, and Why?
Yes, my teaching methods have evolved over the years. And this is not only related to the motivation and context-driven approach, but also using software in the process of understanding what the problem at hands means, and in accumulating intuition about how to approach it (doing particular examples). I also listen to what the students think and their arguments and ideas. I do this much more today than I did years ago. That's because I want to make them participate in the discovery of a solution and also because I learn interesting ideas from them!

Why Is a Collaborative Problem-Based Approach Worthwhile?
I understand this question as, "Is a collaborative problem-solving approach worthwhile?" My answer is yes, it is worthwhile. But it has to be used cautiously! I always encourage collaboration during the math circle I run. But I am usually the person in charge of these, and I actively steer the discussion in directions that are promising. In particular, I listen to all ideas proposed by the students during the discussion and draw their attention to the ones that are interesting and useful. If there are no ideas, I suggest something to help the students move off an idle point.

On the other hand, I have observed that when the students are in charge of such discussions, and the groups of participants are too big, then the effectiveness of the discussion is very low. Students do not listen to each other, insist on their ideas being listened to, and even some of them become lazy and do not work at all, waiting for somebody else to do the job. So, collaborative work is important, but it needs to either be supervised or in reasonably sized groups.

What Are Your Favorite Problems, and Why?

I have many favorite problems. What attracts me to a problem may be the way it is formulated or the witty and surprising idea used to solve it. Most often, I am attracted by the "standing" of the problem in the body of mathematics involved. The best situation is when the problem needs knowledge from a different area of math. Below are problems that fit these categories.

Example 1. In a triangle, the nine-point circle is tangent to all circles tangent to the side lines of that triangle. (This one reveals the relationship between different attributes of a triangle in the Euclidean plane.)

Example 2. Given a sequence of pairwise distinct real numbers $a_1, a_2, \ldots, a_m$, if $m > n^2$, then there is a strictly monotone sub-sequence of length $n+1$. (This is a problem with a beautiful and unexpected idea in the proof.)

Example 3. If the plane figure F has diameter d, then there is a regular hexagon of height d that contains F. (This can be proved by using the analysis concept of continuity and the geometric properties of plane figures.)

Vlad Crisan, PhD Student in Mathematics at University of Goettingen in Germany, with Bachelor's and Master's Degrees in Mathematics from the University of Cambridge in the United Kingdom

What Is Your Personal Approach to Teaching Problem Solving? How Do You Ensure the Kids Are Learning and the Process Is Effective?

When I prepare for a class at AwesomeMath, I typically start with a core of the theory for the topic I would like to teach and then proceed in two recursive steps:

- I first select the problems and make sure that they cover as many facets of that topic as possible.
- I then spend a lot of time thinking how to present the theory and examples so that the context of the topic is explained very clearly and all the solutions to the problems we cover feel natural and not like an "out-of-nowhere" trick.

I may then repeat these "adjust the problems – improve the theory" steps a few times until I feel I have good material for the class. As a result, the students are presented with a refined overview of the theory and also with a large subset of the key strategies for that topic, both of which are typically acquired with years of experience.

I believe it is the teachers' role to pave a good path for any direction the student would like to take to deepen their exploration of a topic and that it is the students' duty to explore as many of those paths as possible through the problems given by the teacher and others.

During class, I encourage a lot of interaction, regardless of whether we are discussing the theory or some problems, with one of the mottos being "there is no such thing as a silly question." To boost the students' problem-solving skills, I challenge them to explain to me how they came up with the key idea in their solution. This approach helps the students stay connected to the class and help me in turn with adjusting my pace. What I also noticed is that the students get a lot of their enthusiasm for a topic from my enthusiasm when teaching it. This is perhaps one of the biggest challenges in general as a teacher, since you cannot fake enthusiasm.

George Catalin Turcas, PhD Student in Mathematics at the University of Warwick in the United Kingdom with a Master's Degree from the University of Cambridge

What Were Your Own School Experiences Like in Your Country That Contributed to Your Love of Problem Solving?

I am from Romania, and my love for problem solving started way before I went to school. My father had a passion for problem solving and a great admiration for students that were topping the Romanian National Math Olympiad and qualifying to the IMO. My earliest memories include some in which my father was reading the newspapers where Romanian students returning from the IMO were interviewed and he was speaking to me about them with great admiration. Although I never set out an objective of being like these students or even becoming a "mathlete," seeing a spark in my father's eyes every time I explained to him how I'd solved a puzzle or a problem definitely had a huge impact on my further interest for mathematics.

Around 1998, when I entered school, the first four years of the mathematical curriculum were filled with boring exercises in which one had to know how to use the order of operations (PEMDAS: Parenthesis, Exponents, Multiplication, Division, Addition, Subtraction) and some unduly called "problems" for which one just had to identify all the numbers in the text, then translate the text into a sequence of operations using those numbers and solve the latter using the same PEMDAS. I suspect nothing has changed for the better since that time. There was not too much thinking going on. My luck was that I never considered that "mathematics." I understood way before I knew how to explain it that mastering computations is necessary for doing mathematics as mastering a language is necessary for sharing ideas with other people. But luckily, I was by then aware that there is a whole mathematical universe out there full of mysteries that can be discovered using rigorous thinking. My father was always there with a more challenging puzzle. Some of my peers were not so lucky.

Nothing much happened at school regarding mathematics education until fifth grade when we were permitted to participate in Olympiads. These competitions are of great interest in Romania, and what I found unique to other countries is their scale. Everybody decent at mathematics is encouraged (sometimes forced) by their teachers to participate in the first rounds of the Romanian Mathematical Olympiad. The continuous Olympiad training soon became my main source of mathematical education. I have many memories of learning mathematical concepts such as the theory of linear systems of equations, induction, or advanced notions such as continuity of functions very early because they would help me understand better a phenomena encountered in competition mathematics. These concepts became so natural to me and, years later, when we were learning about them in school or university, I was able to come up with alternative ways of explaining them to my fellow students.

Because I've learned a lot of mathematical concepts through problem solving, I've never had to ask myself, "Where are we going to use this?" It was already clear that most concepts I'm learning explain something that I previously thought is a surprising phenomenon, but which can be very well explained using some more advanced concepts or theories. Getting to the core of that explanation and understanding that the "surprising phenomena" shouldn't be very surprising was and still is extremely satisfying for me.

What Is Your Personal Approach to Teaching Problem Solving? How Do You Ensure the Kids Are Learning and the Process Is Effective?
It goes without saying that an effective instructor must spend a decent amount of time preparing the examples and problems he chooses to discuss. While presenting the examples, I'm trying my best to show them how to think. For the latter, I have a few different approaches. When possible, I choose to dismiss some part of the hypothesis and show that the conclusion we initially expected does not hold anymore. This sometimes helps them realize that in order to solve the initial problem, they must exploit somehow the part of the hypothesis I dismissed at the previous step.

Other times, if the initial question asks them to prove that statement A implies statement D, in my head I break the problem into a chain of implications, something like A implies B implies C implies D. I might ask them to think how they would prove the easier problem "A implies B." After they manage that, I put forward the challenge of proving that statement C implies statement D. By the time they managed to prove that, they will be already working to complete the chain by proving C implies D and therefore finishing the problem. After we have at least one solution for a problem, I like to ask the class what they think was the most important idea used in the proof. By thinking about it, they internalize the fundamental concepts much better. It is also valuable to think about generalizations of the problems we managed to solve together.

It is very important to give them loads of tests and feedback on their performance in tests. With the students I work with more frequently, I do not care too much about their grades in the tests, because they might vary a lot in difficulty from one week to another. I also don't want the students to think about what they're going to score with the time they should be spending thinking about problems. Nevertheless, I do pay a lot of attention to and keep track of the feedback forms that I write for them.

How Have Your Teaching Methods Evolved Over Time, and Why?

In my first years of teaching, I used to prepare way too much material for each lesson. Although I was constantly trying not to rush and most students could follow my explanations, I was showing too many examples and not leaving enough time for them to think. It took me a while to understand that there is a difference between students being able to follow me as I go step-by-step from first to last and them being able to produce a similar long sequence of steps in order to solve a slightly different problem. By showing them too many examples and too many solutions in class, I was not giving them the opportunity to learn how to discover such a long sequence of steps. Now, I spend more time picking one or two of what I think are instructive examples and even more time choosing the right problems for them to think about. The latter is much harder than it seems. Thinking about problems is very time consuming, and I try my best to give them the problems for which the process of discovering the solution is the most enlightening.

Dr. Branislav Kisacanin, Computer Scientist at Nvidia Corporation

What Were Your Own School Experiences Like in Your Country That Contributed to Your Love of Problem Solving?

When I was growing up in former Yugoslavia, during the 1980s, math and physics competitions were well organized and students were encouraged to participate. Competitions were held at school, city, regional, and national level, and from there teams were sent to the IMO and the IPhO. Except for the school level, all competitions involved some kind of travel with like-minded kids, and that was a big part of it all for me. Thanks to these competitions, I met many life-long friends (my fellow students and my future college professors) and visited wonderful places in former Yugoslavia: Postojna cave and Portoroz in Slovenia, Sarajevo in Bosnia, Decani, with its famous fourteenth century monastery, and the Danube's Djerdap Gorge in Serbia.

What Is Your Personal Approach to Teaching Problem Solving? How Do You Ensure the Kids Are Learning and the Process Is Effective?

Someone wise said that the best teachers and leaders do not lecture or command, they inspire. That is my own experience, too: Students learn best when they are inspired by their own curiosity. So, the question becomes "How do we inspire curiosity?" My approach to it has several facets:

- Tell students about famous mathematicians and scientists and their contributions. Archimedes, Newton, Euler, Einstein, and Tesla are sure to fire up their imaginations!
- Show them formulas and theorems that they have not seen in the regular school with the excuse of these concepts being "very advanced." I turn that upside down and show them things like the beautiful consequence of Euler's formula, $e^{i\pi} + 1 = 0$, because it is the students who are "very advanced," not concepts.
- Show them multiple approaches to solving a problem or proving a theorem. I will always remember when I got an applause from a group of fifth and sixth graders at the Metroplex Math Circle for showing them three very different ways to prove that $\binom{n}{0} + \binom{n}{1} + \binom{n}{2} + \ldots + \binom{n}{n} = 2^n$, combinatorial, by induction, and using Newton's binomial formula.

How Have Your Teaching Methods Evolved Over Time, and Why?

Over the years, I became increasingly confident that inspiring students is more important than lecturing, as is letting the students present their solutions in front of their peers, instead of me presenting all problems. Furthermore, having students present, improves their confidence and presentation skills, so critical in the modern world. When I started working with mathematically gifted kids, I would spend 20% of time on inspirational stories, 0% of time letting students present, and 80% on lecturing. Now, 15 years later, the shares are closer to 40%, 30%, and 30% and continuing to evolve in that direction.

Why Is a Collaborative Problem-Based Approach Worthwhile?

The social aspect of collaborative problem solving is immeasurably beneficial for developing young minds. You cannot put a price or cost/benefit number on:

- Knowing that they are not alone in their love of knowledge and curiosity.
- Using positive peer pressure (yes, there is such a thing, even though we only hear about the negative peer pressure).
- Making life-long friends they will encounter again in college, scientific conferences, and when they travel to Stockholm to pick up their Nobel prize.

What Are Your Favorite Problems, and Why?

My favorite problems are the problems that in the end require two or three lines to solve, but you first need to spend 10–15 minutes thinking about where to even start. My favorite example for the super-smart fifth and sixth graders is the problem asking to find all positive integers $1! + 2! + \ldots + m! = n^2$.

Waldemar Pompe, University of Warsaw in Poland, Institute of Mathematics

What Were Your Own School Experiences Like in Your Country That Contributed to Your Love of Problem Solving?

Growing up in Poland, it was my mathematics teacher who inspired me. I was bored and would then be disruptive during the class, so the teacher gave me more problems to solve, although they weren't very difficult. However, at the end of the booklet that was provided, I found more interesting and complicated problems, and I was excited to try and solve them. Later, I discovered that the math Olympiad exists with even more of these types of problems.

What Is Your Personal Approach to Teaching Problem Solving? How Do You Ensure the Kids Are Learning and the Process Is Effective?

I'm never sure that the process is effective. I'm trying to select problems that are neither too easy nor too hard. When you teach in front of the students, you usually feel, if you should select something easier or more difficult for the next time. What is important for me, I never think about "training" for the Olympiad. The most important goal is that students have the joy of solving problems and want to concentrate on mathematics.

What Are Your Favorite Problems, and Why?

Those that have an unexpected answer and a clever explanation.

Dr. Oleg Mushkarov, Bulgarian Academy of Science; Institute of Mathematics and Informatics

When the road to the inn is more interesting than the inn...

What Are Your Favorite Problems, and Why?

As an example, I would like to discuss a problem I love to share with my undergraduate students in the frames of a course on extracurricular math activities. Unlike the traditional situations, when I expect the students to come to a solution (possibly with my help), I present a solution containing mistakes to a well-defined problem and throw the gauntlet down for them: *Find the mistake in this solution.*

These types of problems are something I experienced myself as a member of the Bulgarian team for the first Balkan Mathematical Olympiad for university students and young researchers held in September 1971 in Bucharest, Romania. One of the problems was to find the mistake in a proposed "solution" to a given problem dealing with abstract algebra. I am convinced that such problems are essential since everyone (even a professional mathematician) makes errors on an everyday basis.

1. What matters first is to feel that something is wrong (to become aware of it).
2. Then discover where the mistake is.
3. Finally, find a correct solution to the problem.

The problem I have in mind is one I tried to solve some 30 years ago and my first attempt to do that turned out to be wrong. It is the very process of starting with a wrong idea, becoming aware of the error in reasoning, and finally, succeeding to correct it, which I have kept reproducing when teaching that course. And it is this process I'll try to share with you now.

To reproduce the experience gained in this context as close as possible, I invited my colleague and friend, Jenny Sendova, to take notes and snapshots while I was demonstrating to her my presentation to the students.

What Jenny Recorded

Here is the problem: *There are four cities located at the vertices of a square. Build the shortest road network (system of paths) connecting the cities.*

And here is how I present my "solution" to the students. We shall use two important mathematical facts:

1. The shortest path between two points on the plane is the segment connecting them (Figure 7.1).
2. If a continuous function has values of opposite sign at the end points of an interval, then it has a root in its interior (Bolzano's theorem [http://mathworld.wolfram.com/BolzanosTheorem.html]) (Figure 7.2).

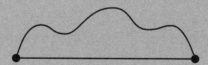

Figure 7.1 The shortest path between two points on the plane is the segment connecting them.

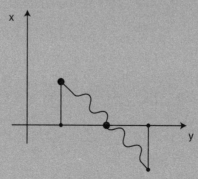

Figure 7.2 Bolzano's theorem says that if a continous function is sometimes positive and sometimes negative, it must be zero as some point.

We can start with an arbitrary system of paths (represented by continuous curves connecting the cities which are the vertices of a square ABCD) (Figure 7.3):

Then there exists a path connecting the cities A and C and a path connecting the cities B and D. Without loss of generality, we may assume that both paths lie in the square; applying the Bolzano's theorem, we know that there exists a point M in which two paths meet (Figure 7.4). The new system still connects the four cities

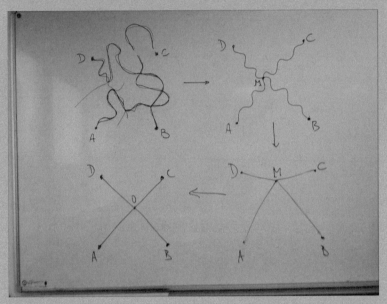

Figure 7.3 Continuous curves connect the cities that are the vertices of a square ABCD.

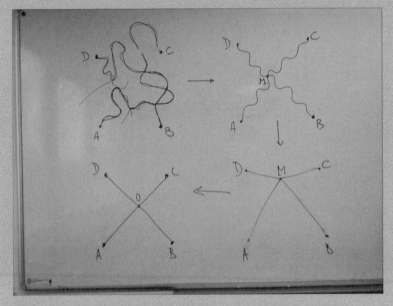

Figure 7.4 According to Bolzano's theorem, there exists a point M in which two paths meet.

and is of a shorter length. Applying (1) we can reduce the paths to segments and thus get an even shorter system of paths, namely consisting of the segments AM, BM, CM, and DM (Figure 7.5). Now, in virtue of the triangle inequality, we can reduce our system of paths to the diagonals of the square (Figure 7.6), which implies that this is the shortest system of paths, right?

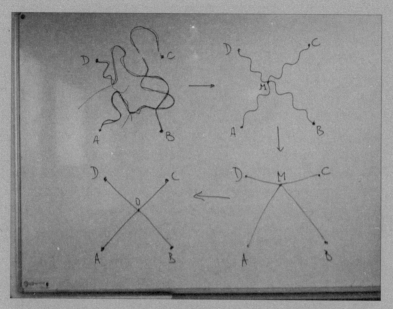

Figure 7.5 A shorter system of paths consists of the segments AM, BM, CM, and DM.

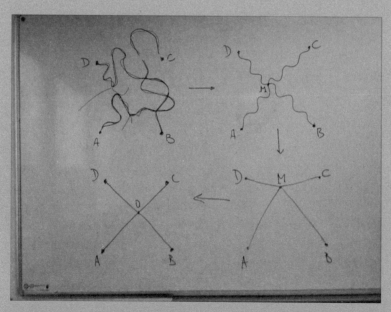

Figure 7.6 The system of paths can be reduced to the diagonals of the square.

Yes, but no!!! (as famous Bulgarian journalist Petko Bocharov used to say). When I came up with this "solution," I felt that something was wrong, since I remembered the *Steiner tree problem* and I had a vague memory of a picture in the great book of Courant and Robbins, *What Is Mathematics*. But instead of digging into the literature, I was eager to find where the mistake (the catch) in my reasoning was – in other words, *where the dog lay buried* (as the Bulgarians would say in this case).

Let us come back to our mathematical facts, namely Bolzano's theorem. Thanks to it we were sure that there is an intersection point between the two paths connecting the opposite vertices of the square. *An* intersection point? Yes, but who says that it is a single one? The precise statement of Bolzano's theorem is that there is *at least* one intersection point, but there may be more! This was the *Aha!* moment for me!

The above observation shows that, in general, the first reduction of the system of paths in Figure 7.4 should not look like the one in Figure 7.5 but rather like the system in Figure 7.7. Hence, considering the first and the last intersection points of the paths connecting the opposite vertices of the square and applying, we come to the problem of minimizing the sum $AM+DM+MN+BN+CN$, where M and N are points inside the square ABCD (Figure 7.8).

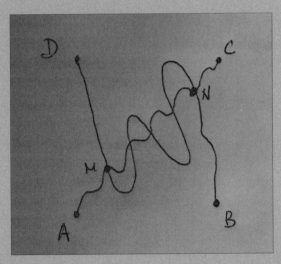

Figure 7.7 The first reduction of the system of paths should look like this system.

Proceeding with Heron's "shortest distance" problem (Figure 7.9), then introducing two parameters (the angles between the path segments and the sides of the square) and using some trigonometry, we get a function of one parameter.

Finding its minimum gives us a genuine solution (Figure 7.10).

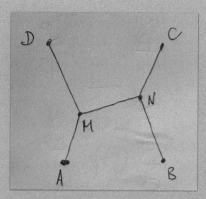

Figure 7.8 Now consider how to minimize the sum $AM + DM + MN + BN + CN$, where M and N are points inside the square $ABCD$.

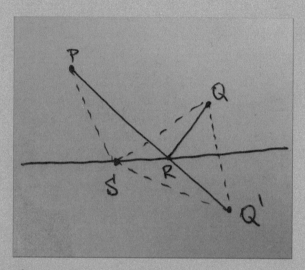

Figure 7.9 Heron's "shortest distance" problem.

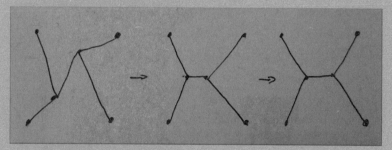

Figure 7.10 Finding the minimum gives us a genuine solution.

Finally, note that the problem has two distinct solutions up to symmetry (Figures 7.11 and 7.12).

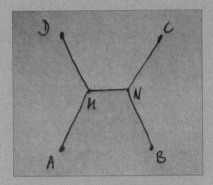

Figure 7.11 One of two possible solutions with symmetry.

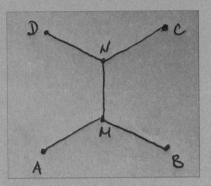

Figure 7.12 A second distinct solution with symmetry.

And here is a short series of snapshots Jenny took during the lecture.

Post Reflections

After the demonstration, Jenny and I decided to dig up some interesting facts about the history of the problem. We found numerous sources. For example, in one we read that the problem of finding the shortest network connecting cities can be traced back to a correspondence between Schumacher and Gauss in 1836. The problem Schumacher was originally interested in – namely, to find a point that connects a given set of points with the shortest network – was even older and goes back to 1638, when Descartes asked Fermat to study curves whose points have a constant sum of distances to four given points.

Motivated by this question, Fermat asked in 1643 for the case of three given points: Which point would minimize the sum of distances? Torricelli was the first to solve the three-point case (the problem became known as the Fermat-Torricelli problem). Fagnano solved the case of four points. It is worth mentioning that while the solution for up to four points could be done by ruler and compass, such a construction is not possible for more points in general position (a result obtained by applying Galois theory). This type of problem is now known as *Steiner tree problem*, after the famous geometer Jacob Steiner (although it is not quite clear what his contributions to this problem have been).

Of course, if I had tackled the problem with first inquiring what was known about it, I might have never gone along this road. This is often a problem with today's students, who would rather "google" for facts before even starting to think about possible solutions. The real problem for me, though, was to find the mistake in my first idea (based not only on intuition but on seemingly well-known mathematical facts).

After sharing the whole experience and presenting a rigorous proof to Jenny, she proposed a more *natural* proof – namely, using a soap film. Indeed, there was such a film on YouTube (far from soap opera, though as she noted [http://www.youtube.com/watch?v=dAyDi1aa40E]).

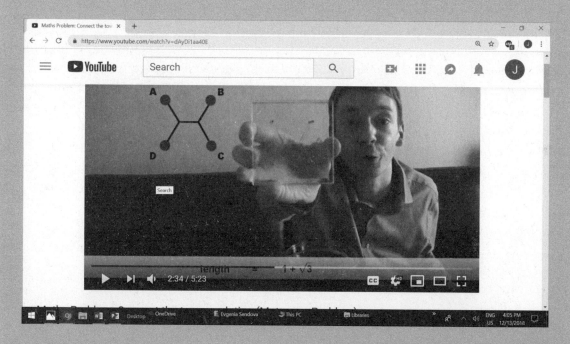

Then we found the picture I was thinking about in the book *What Is Mathematics*:

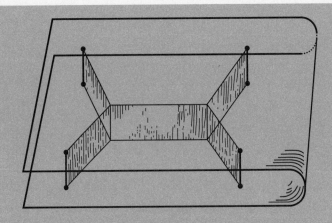

How lucky I was to have forgotten this picture, but not – *what mathematics is*! Probably, the experimental physicist Arthur Stanley Eddington is right to think that *proof is an idol before whom the pure mathematician tortures himself* – but they forget to add that it is the tortures and the successful debugging process that let us experience the pure joy at the end.[8]

Further Reading

Gander, M.J., Santugini, K., and Steiner, A. (2008). Shortest road network connecting cities. *Bollettino dei docent di matematica* 56: 9–19. http://www.unige.ch/~gander/Preprints/BDM56-GanderE.pdf.

Chen, X., and Du, D.-Z. (2002). *Steiner Trees in Industry*, Handbook of Combinatorial Optimization, 5e (eds. D.-Z. Du and P.M. Pardolos). Kluwer Academic Publishers http://citeseerx.ist.psu.edu/viewdoc/download?doi=10.1.1.122.3025&rep=rep1&type=pdf.

Brazil, M., Thomas, D., Graham, R., and Zachariasen, M. (2014). On the history of the Euclidean Steiner tree problem. *Archive for History of Exact Sciences* 68 (3): 327–354. http://www.researchgate.net/publication/262300152_On_the_history_of_the_Euclidean_Steiner_tree_problem.

Du, D.Z., Hwong, F.K., Song, G.D., and Ting, G.Y. (1987). *Steiner Minimal Trees on Sets of Fourt Points*, Discrete and Computational Geometry, 2e, 401–414. New York: Springer-Verlag.

Du, D.Z., Yao, E.Y., and Hwang, F.K. (1982). A short proof of a result of Pollak on Steiner minimal trees. *Journal of Combinatorial Theory. Series A* 32: 396–400. https://link.springer.com/content/pdf/10.1007/BF02187892.pdf.

Notes

1. http://www.cut-the-knot.org
2. http://nlvm.usu.edu/en/nav/vlibrary.html
3. https://wordplay.blogs.nytimes.com/2016/07/18/imo-2016
4. https://artofproblemsolving.com/community
5. National Association of Math Circles, http://www.mathcircles.org/find
6. Sam Vandervelde, *Circle in a Box* (Mathematical Sciences Research Institute, 2007), http://www.mathcircles.org/wp-content/uploads/2017/07/circleinabox.pdf.
7. Math Teachers' Circle Network, www.mathteacherscircle.org.
8. Arthur Eddington, *The Nature of the Physical World: Gifford Lectures (1927)* (Cambridge, UK: Cambridge University Press, 2012), Originally published in 1927.

CHAPTER 8

Mini-Units

These mini-units are printables that you can use for a 15–20-minute problem-based learning (PBL) lesson. Please note that some of the lessons can be split into 10-minute units given over a two-day period. The units provide an opportunity to:

- Ease into a PBL environment with manageable content.
- Have access to short lessons on days when you have other pacing requirements to satisfy.
- Make good use of transition times at the start or end of classes to capture attention or provide a fun mental break with a collaborative challenge.

When going through the mini- and full units, it is important to take advantage of the Cornell note-taking method discussed in Chapter 6 through the *relate, reflect, and revise* process. That way, the units will evolve with you and your students. The following questions will help you and your students process what has been covered and get the most out of the units.

Relate/Reflect/Revise Questions

Teachers

- Were students participating and engaged?
- Did they understand how the lesson relates to other areas of mathematics?
- Do you need to reteach? If so, what parts?
- What feedback did you get from the students?
- What would you do differently?
- What additional resources could you use?

Students

- How would you use this lesson's knowledge moving forward?
- What areas were difficult to understand?
- Did you ask for help from peers or the teacher?
- If you were teaching the lesson, what would you do differently?

Roman Numeral Problems

OVERVIEW

The Roman numeral system played an important role in the era of mathematics. The Romans initially developed this counting system for commercial purposes by using the Latin alphabet to represent numerical values. The Europeans continued to utilize this system until the 1600s. People have incorporated Roman numerals in artifacts, inscriptions on buildings, clocks, chapters in books, and chemistry as well. Therefore, the study of Roman numerals helps people to better analyze and understand the workings of the world. This allows students to grow their critical thinking skills through exposure to practical and interesting problems.

LEARNING OBJECTIVES

By the end of the lesson, students will know basic operations with Roman numerals and be able to convert years from Roman numerals to Arabic numerals and the other way around. Why is this important? Because they provide a good example of a nonpositional number system different from today's decimal system.

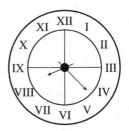

SUGGESTED LESSON PLAN

1. At the beginning of the class you can have the following two problems on the board:
   Can anybody recognize the years?
   Leonardo da Vinci was born in MCDLII, and died in MDXIX.
   Marilyn Monroe was born in MCMXXVI, and died in MCMLXII.
2. Explain the rules.
3. Have fun with problems: group and independent problems.
   Roman numerals in modern day numbers

   I V X L C D M

   M = 1000

   D = 500

   C = 100

L = 50
X = 10
V = 5
I = 1

How to represent numbers that are less with 1 unit than V and X:

4 = 5 − 1 = IV
9 = 10 − 1 = IX

How to represent numbers that are less with 10 units than L and C:

40 = 50 − 10 = XL
90 = 100 − 10 = XC

How to represent numbers that are less with 100 units than D and M:

400 = 500 − 100 = CD
900 = 1000 − 100 = CM

OBSERVATION

In order to subtract one number from another in the Roman numeral system, you must place no more than one lesser value to the left of the greater number from which you are subtracting.

How are the other numbers represented?
The rest of the numbers are obtained by addition.

2 = 1 + 1 = II
3 = 1 + 1 + 1 = III
8 = 5 + 3 = VIII
14 = 10 + 4 = XIV
20 = 10 + 10 = XX

Solve

76 = 50 + 20 + 6 =

Repeating a *numeral* up to three times represents addition of the number. Only I, X, C, and M can be repeated; V, L, and D cannot be, and there is no need to do so.

PROBLEMS

1. How do you write 2019?
   Solution: 2019 = 1000 + 1000 + 10 + 9 = MMXIX

2. Group Activity:
   Have three students go to the board and compete on who is the fastest writing the following numbers as Roman numerals:
   35, 71, 18, 46, 251, 859, 2297, 3160
   Solution:
   35 = 30 + 5 = XXXV
   71 = 50 + 20 + 1 = LXXI
   18 = 10 + 5 + 3 = XVIII
   46 = 40 + 5 + 1 = XLVI
   261 = 200 + 50 + 10 + 1 = CCLXI
   859 = 500 + 300 + 50 + 9 = DCCCLIX
   2297 = 2000 + 200 + 90 + 7 = MMCCXCVII

3. Independent Activity

   Easy
   Write the year you were born in Roman numerals.
   Write the following Roman numeral numbers in Arabic numbers:
   LXII, XV, LXXXIX, CCLVII, DVIII, DLXXXVI, MDCXXXIII, MMVII, MCMXCVIII
   Answers: 62, 15, 89, 257, 508, 586, 1633, 2007, 1998

   Challenging
   Let N be the greatest positive integer that can be expressed using all seven Roman numerals I, V, X, L, C, D, and M, exactly once and let n, be the least positive integer that can be expressed using these numerals exactly once each. Find $N - n$. Note that the arrangement CM is never used in a number along the numeral D.
   (Purple Comet Math Meet Contest 2019, created by Titu Andreescu)

   Solution: 222
   The number N is expressed by listing the numerals in decreasing order of value: MDCLXVI, which represents 1666. The least is obtained by writing the smallest valued numerals before the larger ones: MCDXLIV, which represents 1444. The requested difference $1666 - 1444 = 222$.

CHECK FOR UNDERSTANDING

Going back to the years presented on the board at the beginning of class, what are the answers?
   Answers: Leonardo da Vinci was born in 1452 and died in 1519. Marilyn Monroe was born in 1926 and died in 1962.

Cryptarithmetic

OVERVIEW

Cryptarithmetic is the art and science of creating and solving cryptarithms. A *cryptarithm* is a genre of mathematical puzzles in which the digits of an arithmetic problem are replaced by letters of the alphabet or other symbols. The invention of Cryptarithmetic has been ascribed to ancient China. In India, during the Middle Ages, arithmetical restorations or *skeletons* were developed, a type of cryptarithms in which most or all of the digits have been replaced by asterisks.

The world's best-known alphametic puzzle is undoubtedly SEND + MORE = MONEY. It was created by H. E. Dudeney and first published in the July 1924 issue of *Strand* magazine.

LEARNING OBJECTIVES

This unit stresses the importance of place value, reinforces reasoning skills, and narrows down the puzzle to a manageable case analysis. It shows students how to look for a good start when solving a problem. After that, the number of cases significantly decreases; the student learns how to rule out cases that are not possible until the unique solution emerges.

FOUNDATIONS

Cryptarithmetic Conventions

1. Each letter represents only one digit throughout the problem.
2. When letters are replaced by their digits, the resulting arithmetical operation must be correct.
3. The numerical base, unless specifically stated, is 10.
4. Numbers may not begin with a zero.

Preparation

Rewrite the problem, expanding the space between the lines, to make more room for trial numbers that will be written underneath the letters, like this:

```
    S E N D
    9 5 6 7
+   M O R E
    1 0 8 5
-----------------------
  M O N E Y
  1 0 6 5 2
```

Example 1

$$\begin{array}{r} AB \\ BC \\ + CA \\ \hline ABC \end{array}$$

Solution
Looking at the ones digits, we see that $A+B=10$. Taking into account the carryover, looking at the tens digit we have $C+1=B$ (we cannot have $C+1=B+10$ because that would force $B=0$, a contradiction). Then $A=1$, $B=9$, and $C=8$.

Example 2

$$\begin{array}{r} MOO \\ + MOO \\ \hline COW \end{array}$$

Solution
Looking at the tens digits, we must have $O+W+1=O+10$, with carryover (O cannot be zero, otherwise W would be zero as well, a contradiction). It follows that $O=9$ and $W=8$. There are three pairs (M, C) for which $2M+1=C$ and C is different from W, namely (1,3), and (2,5), and (3,7). Hence there are three solutions to the problem.

Problems
Solve the following problems[1]:

Easy

1. $$\begin{array}{r} XX \\ YY \\ + ZZ \\ \hline XYZ \end{array}$$

2.
```
      T E A C H
  +     M A T H
  -------------------
      G I F T E D
```

Solution

1. Looking at the ones column, we see that the ones digit of X + Y + Z is Z. Therefore, the ones digit of X + Y is 0, which is only possible when X + Y = 10. This means a 1 is carried over to the tens column. Looking at the tens column, the ones digit of X + Y + Z + 1 is Y. Similarly, this means X + Z = 9 and a 1 is carried over to the hundreds column. Looking at the hundreds column, we get X = 1. Since X + Z = 9, that means Z = 8. Since X + Y = 10, that means Y = 9. This gives the desired solution. Another way of looking at this is to use an equation:

$$10X + X + 10Y + Y + 10Z + Z = 100X + 10Y + Z$$

   implying that Y + 9Z = 89X.

   It thus follows that X = 1, Y = 8, and Z = 1.

2. Because GIFTED is a six-digit number, TEACH is a five-digit number, and MATH is a four-digit number, we must have T = 9 so that the sum is large enough. The largest possible value of TEACH + MATH is 99 999 + 9999 = 109 998, so we must have G = 1, I = 0. Also, in one of the columns, two A's add up to T = 9. The only way this can happen is if A = 4, and a 1 is carried over from the previous column during addition. In the rightmost column, two H's add up to D. H can only be 3, 6, or 8, because otherwise either H or D will take a value which has already appeared before. If we try H = 3, then D = 6, and the only possible value for C is C = 8. Hence E = 7, the remaining digits are 2 and 5, and we can assign M = 5 and F = 2. This gives the desired solution.

```
        97 483
    +    5 493
    ---------------
       102 976
```

Challenging

3. In the subtraction PURPLE − COMET = MEET, each distinct letter represents a distinct decimal digit, and no leading digit is 0. Find the number PURPLE.[2]

   Answer: 103 184

Write the problem as an addition problem in the following form:

$$\begin{array}{r} \text{COMET} \\ + \text{MEET} \\ \hline \text{PURPLE} \end{array}$$

Solution

Clearly, the P digit results from a carry of 1, so P represents 1. Then it must be that $C+1=U+10$, and it follows that U represents 0 and C represents 9. The value of T determines E, the value of E determines l, the value of L determines M, and the values of O and R can then be determined to form a correct addition problem. The following table indicates the possible interferences.

T	E	L	M	O	R
2	4	8	7	5	3
3	6	2	4		
4	8	6	4		
5					
6	2	5			
7	4				
8	6	3	2		

The only possible solution is for T to represent 2, E to represent 4, L to represent 8, M to represent 7, O to represent 5, and R to represent 3. The value of PURPLE is then 103 184.

4.

$$\begin{array}{r} \text{FORTY} \\ \text{TEN} \\ + \text{TEN} \\ \hline \text{SIXTY} \end{array}$$

Solution

From the ones column, we must have $Y+2N=Y$ or $Y+10$; hence, $N=0$ or 5. If $N=5$, then we have a carry of one out of the ones column, and the tens column gives $T+2E+1=T$, $T+10$, or $T+20$, all of which are impossible. Thus, we must have $N=0$ and $E=5$. Note that this gives a carry of one out of the tens column. Since I is at least 1, we have $SI-FO \geq 2$.

Thus, we must have a carry of two out of the hundreds column. Further, this still requires $I = 1$, and $O = 9$. Since we have determined all the carries, the remaining columns give us $R + 2T = X + 19$, and $S = F + 1$. The first equation requires R and T to be both among 6, 7, and 8, and X to be at most 4. Hence, F and S being consecutive must be 2 and 3, or 3 and 4. In particular, one of them is 3, so X must be 2 or 4. Thus, the equation $R + 2T = X + 19$ shows that R is odd, hence $R = 7$ and $2T = X + 12$. Hence, $T = 8$ and $X = 4$. Thus, $F = 2$ and $S = 3$.[3]

$$
\begin{array}{r}
29{,}786 \\
850 \\
+850 \\
\hline
31{,}486
\end{array}
$$

Squaring Numbers: Mental Mathematics

OVERVIEW

Have you heard of Dr. Arthur Benjamin, the "mathemagician" of our times? He is a professor of mathematics at Harvey Mudd College and is known for mental math capabilities and "Mathemagics" performances in front of live audiences. Imagine being able to accomplish calculations faster than a calculator! Dr. Benjamin has unlocked the Secrets of Mental Math and performs complex calculations in his head with ease. Throughout this chapter you will be exposed to one of his secrets: You can be a mathemagician, too!

Do unto one side as you would do unto the other.

Arthur Benjamin

LEARNING OBJECTIVES

By the end of this unit, you will be able to easily square two-digit numbers in your head without the use of a calculator. This is helpful for students who wish to apply their algebraic knowledge and see how mental computations can be simplified when using algebraic identities.

You will need knowledge of basic arithmetic and algebraic factoring.

PROBLEMS

We present two examples of how we can use these identities for quick "mental" squaring of numbers (without a calculator).

1. The first trick relies on the identity:

   $$n^2 = (n-k)(n+k) + k^2$$

   We choose k conveniently, to get to the closest multiple of 10 or 100.

   Examples:
   $34^2 = 30 \times 38 + 4^2 = 1156$
   $48^2 = 46 \times 50 + 2^2 = 2304$
   $75^2 = 70 \times 80 + 5^2 = 5625$

   Solve, using the above method:
   57^2
   63^2
   89^2
   113^2

   Solutions
   $57^2 = 54 \times 60 + 3^2 = 3249$
   $63^2 = 60 \times 66 + 3^2 = 3969$
   $89^2 = 78 \times 100 + 11^2 = 7921$
   $113^2 = 100 \times 126 + 13^2 = 12\,769$

2. The second method used for fast squaring of numbers 30 through 80 is based on the following identity:

   $$n^2 = 100 \times (n+25)^2 + (n-50)^2.$$

   $34 = 9 \times 100 + (-16)^2 = 1156$
   $48^2 = 23 \times 100 + (-2)^2 = 2304$
   $75^2 = 50 \times 100 + 25^2 = 5625$

   Solve using the second method:
   39^2
   57^2
   63^2
   78^2

   Solutions
   $39^2 = 14 \times 100 + (-11)^2 = 1521$
   $57^2 = 32 \times 100 + 7^2 = 3249$
   $63^2 = 38 \times 100 + 13^2 = 3969$
   $78^2 = 53 \times 100 + 28^2 = 6084$

The Number of Elements of a Finite Set

OVERVIEW

We often need to count the number of terms in a succession. This unit presents some examples that illustrate this idea. It gives students a flavor of some introductory counting techniques.

LEARNING OBJECTIVES

This unit helps develop the counting skills of students. *Combinatorics* is the art of counting minus the counting! This topic introduces the "art" by considering elementary cases dealing with finite sets.

SET NOTATION

{}	set, a collection of elements; for example A = {1, 2, 4, 6},
|	such that, for example, A = {$x \mid x \in R, x > 0$}
A∩B	intersection, objects that belong to set A and B
A∪B	union, objects that belong to set A or B
A⊆B	subset, A is a subset of B; set A is included in set B
A⊂B	proper subset/strict subset; A is a subset of B, but A is not equal to B
A⊄B	not subset; set A is not a subset of set B
A = B	equality; both sets have the same members
A\B or A-B	relative complement, objects that belong to A and not to B
a∈A	element of/belongs to
|A|	cardinality, the number of elements of set A
Ø	empty set, Ø = {}

Let $A = \{a_1, a_2, \ldots, a_n\}$ be a finite set. The number of elements of A, in our case n, is called the cardinal number of A. This number is denoted by $|A|$.

The basic properties of the cardinal number are the following:

1. If the sets A and B are disjoint, that is $A \cap B = \emptyset$, then $|A \cup B| = |A| + |B|$.
2. *Inclusion-exclusion principle for two sets*. The following relation holds:
$$|A \cup B| = |A| + |B| - |A \cap B|$$

The argument for the proof of this relation is very simple: When we count the elements of $A \cup B$, we count two times the elements of $A \cap B$; hence, we have to exclude once these elements.

3. *Inclusion-exclusion principle for three sets.* The following relation holds:

$$|A \cup B \cup C| = |A| + |B| + |C| - |A \cap B| - |B \cap C| - |C \cap A| + |A \cap B \cap C|$$

The argument is similar to those presented above for two sets: Exclude the elements of intersections $A \cap B$, $B \cap C$, and $C \cap A$ and then include the elements of $A \cap B \cap C$.

Example 3

How many integers from 1 to 500 are divisible by 3 or by 7?

Solution

Let A be the set consisting in all integers from 1 to 500 that are divisible by 3. Consider B, the set of all integers from 1 to 500, divisible by 7. We have to find $|A \cup B|$. We write A and B as follows:

$$A = \{3k \mid 1 \leq 3k \leq 500\}$$

$$B = \{7l \mid 1 \leq 7l \leq 500\}$$

In order to calculate $|A|$ we need to find the largest k such that $3k \leq 500$. This is 166; hence, $|A| = 166$. In similar way, for $|B|$ we need to find the largest integer l with $7l \leq 500$, and we get $|B| = 71$.

Applying the inclusion-exclusion principle for two sets, we have

$$|A \cup B| = |A| + |B| - |A \cap B| = 166 + 71 - |A \cap B| = 237 - |A \cap B|$$

The set $A \cap B$ consists in all integers from 1 to 500 that are divisible by 3 and by 7; hence, we can write $A \cap B = \{21s \mid 1 \leq 21s \leq 500\}$.

The largest s with property $21s \leq 500$ is 23; that is, $|A \cup B| = 23$. Finally, it follows

$$|A \cup B| = 237 - |A \cap B| = 237 - 23 = 214.$$

Problems

1. Consider the set $A = \{a \mid a$ is a positive integer less than 2009 and $3 \mid a\}$. Find $|A|$.
2. Consider the set $A = \{x \mid x$ is a positive integer and $345 < 3x < 3210\}$. Find $|A|$.
3. Let A and B be two sets such that $|A \cup B| = 20$, $|A| = 16$, and $|B| = 17$. Find $|A \cap B|$.
4. Consider the set $A = \{a \mid a$ is positive integer less than 2009 and $7 \mid a\}$. Find $|A|$.
5. Let $A = \{x \mid x$ is positive integer, $104 < x < 1300$, and $11 \mid x\}$. Find $|A|$.
6. The students in class participate in at least one of the following activities: chess and court tennis. If 15 students play chess, 16 students play court tennis, and 10 participate in both these activities, then find the number of students in this class.

7. How many integers from 1 to 2009 are divisible by 5, by 7, or by 9?
8. How many integers from 1001 to 2000, inclusively, are divisible by 2 or by 3 or by 5?

Solutions

The largest a with the property $7a \leq 2009$ is 287; hence, $|A| = 287$.

5. The smallest x such that $104 < 11x$ is 10. The largest k such that $11x < 1300$ is 118; hence, we can describe the set A as $A = \{11x \mid x = 10, 11, \ldots, 118\}$. We have $|A| = 118 - 9 = 109$.

6. Let A be the set of students playing chess and let B be the set of students playing court tennis. Then we have $|A| = 15$, $|B| = 16$, and $|A \cap B| = 10$. From the inclusion-exclusion principle, it follows that $|A \cup B| = |A| + |B| - |A \cap B| = 15 + 16 - 10 = 21$; hence, the number of students in class is 21.

7. Let A be the set of integers from 1 to 2009 that are divisible by 5, B that are divisible by 7, and C that are divisible by 9. We have $5k \leq 2009$ is equivalent to $k \leq 401$; hence, $|A| = 401$. Also, $7k \leq 2009$ is equivalent to $k \leq 287$, so $|B| = 281$. From $9k \leq 2009$, we get $k \leq 223$; that is, $|C| = 223$. In order to find $|A \cap B|$, observe that $35k \leq 2009$ is equivalent to $k \leq 57$. Also, $45k \leq 2009$ gives $k \leq 44$, and $63k \leq 2009$ gives $k \leq 31$. In this way, we obtain $|A \cap B| = 57$, $|A \cap C| = 44$, and $|B \cap C| = 31$. In order to find $|A \cup B \cup C|$, observe that $5 \times 7 \times 9k \leq 2009$ is equivalent to $k \leq 6$; hence, $|A \cup B \cup C| = 6$. Applying the inclusion-exclusion principle for three sets, we get

$$|A \cup B \cup C| = |A| + |B| + |C| - |A \cap B| - |A \cap C| - |B \cap C| + |A \cap B \cap C|$$
$$= 401 + 281 + 223 - 57 - 44 - 31 + 6 = 769.$$

8. There are 500 divisible by 2, 333 divisible by 3, and 200 divisible by 5. Out of the given numbers, 167 are divisible by $2 \times 3 = 6$, 100 by $2 \times 5 = 10$, and 67 by $3 \times 5 = 15$. Since 33 of the numbers are divisible by $2 \times 3 \times 5 = 30$, applying the inclusion-exclusion principle we get

$$500 + 333 + 200 - 167 - 100 - 67 + 30 = 729.$$

Magic Squares

OVERVIEW

Magic squares are square grids with a special arrangement of numbers in them. These numbers follow a property, such as every row, column, and diagonal adds up to the same number. Their story begins 4000 years ago in China, where, according to legend, a turtle crept out of the Yellow River. The reptile is said to have had dots on its underside positioned in such a way as to make the 3 × 3 square described above. Renaissance astrologers equated them with planets. Well-known mathematicians have studied magic squares: Leonhard Euler, Édouard Lucas, and Arthur Cayley. In the United States, founding father Benjamin Franklin also spent time constructing interesting variations. He created a 16 × 16 square that astounded others.

> *I make no question but you will readily allow this square of 16 to be the most magically magical of any magic square ever made by any magician.*
>
> Ben Franklin

LEARNING OBJECTIVES

To see the beauty of magic squares, develop critical thinking skills and have fun by solving these puzzles and inspiring students to create their own magic squares. Further, magic squares are a good introduction to mathematical constructions, providing an accessible example utilizing arithmetic to create these constructions.

Take a look at this table:

2	7	6
9	5	1
4	3	8

Notice that if we add the numbers in each row, column, and diagonal, we get the same number. This number is 15.

One can say that a magic square of size n consists of the numbers $1, 2, 3, \ldots, n^2$. These numbers are arranged in such a way that every field contains a different number and that the sums of the numbers in each row, column, and diagonal are equal.

Example: Complete the following magic square:

4		2
	1	

For this example one can easily find the number in the middle of the top row, then the number in the middle of the square, and so on.

The above magic square is known as the Lo Shu square. It uses the numbers 1, 2, 3, 4, 5, 6, 7, 8, 9. As it happens, the only magic square that we can make using those numbers is the one we talked about in the first figure. But the magic square in this example looks different. If we look closer, we can notice that the second magic square is exactly the same as the first one but rotated 90°.

Example 4

How about this magic square?

23	28	21
22	24	26
27	20	25

How much are the sums in each row, column, and diagonal? Are they equal?

Yes, they are equal, and the sum is 72. If one looks closer, this is actually a Lo Shu magic square with 19 added to each number in the square.

Easy

In a magic square, the sum of the three entries in each row, column, or diagonal has the same value. The figure shows four of the entries of a magic square. What is x?

		3
x	4	5

Solution

The sum of the entries of the diagonal containing x and 3 must be equal to the sum of the entries of the bottom row; hence, the center entry of the magic square is 6, because $x + 6 + 3 = x + 4 + 5$. Let the leftmost entry of the first row be y. Then the sum of the entries on the diagonal that contains 5 is $y + 6 + 5$. From $y + 6 + 5 = x + 4 + 5$ it follows that $y = x - 2$.

Let z be the center entry of the first row. Then, writing that the sum of the entries of the first row is equal to the sum of the entries of the second column, we get $(x - 2) + z + 3 = 4 + 6 + z$, implying $x = 9$.

The magic square is:

7	8	3
2	6	10
9	4	5

Solve the magic square:[4]

1	8	11	14
12			7
6			9
15	10	5	4

Solution:

1	8	11	14
12	13	2	7
6	3	16	9
15	10	5	4

Solve the magic square:[5]

1	8	10	15
	13	3	
7			
14			4

Solution:

1	8	10	15
12	13	3	6
7	2	16	9
14	11	5	4

Challenging

An $n \times n$ magic square is filled with the numbers $1, 2, \ldots, n^2$ such that the sum of the entries on each row, each column, and each of the diagonals is the same. If for some $n > 6$ we remove the number 37 from this square, the sum of all other entries in its row is 2019. Find n.[6]

Solution

The sum of all entries in the magic square (including the number 37) is

$$1 + 2 + \ldots + n^2 = (n^2(n^2+1))/2.$$

so the sum of the entries in each row is

$$(1 + 2 + \ldots + n^2)/n = n(n^2+1)/2.$$

Then

$$n(n^2+1)/2 - 37 = 2019.$$

implying

$$n(n^2+1) = 2 \times 2056 = 16 \times 257.$$

It follows that $n = 16$.

Toothpicks Math

OVERVIEW

For many years, scientists tried to measure the benefits of puzzles on the human mind. Almost all researchers agree that puzzles help cognitive processes. In addition, researchers find that doing puzzles daily tends to be most effective. The main psychological benefits of doing puzzles on a regular basis are that the process improves memory, develops creativity, and teaches students new ways of thinking about other everyday problems. Usually there is more than one way to solve a toothpick math problem. A trial-and-error method can sometimes solve it. Some people can immediately visualize the solution, and then they work backwards to see how they did it. You can have your students talk out loud about how they are thinking about the puzzle, or to tell you what steps they took if they managed to solve the puzzle quickly.

LEARNING OBJECTIVES

Puzzles are useful to improve logic and geometric intuition by developing students' reasoning skills as well as their understanding of geometric figures and their properties.

FOUNDATIONS

Manipulatives. A box or two of toothpicks (depending on the class size)

Prior knowledge. Definition of congruent squares

Suggestion. It is more fun if students are divided in groups and they work together in solving the puzzles.

PROBLEMS

There are many interesting problems with toothpicks. Here is a look at some of them:[7]

1. Move two toothpicks in such a way as to obtain five congruent squares.

   Solution:

2. Remove eight toothpicks such that the remaining ones form four congruent squares.

   Solution:

3. Remove four toothpicks to obtain five congruent squares.

   Solution:

4. Remove six toothpicks as to obtain three squares.

Solution:

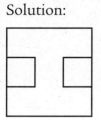

Pick's Theorem

OVERVIEW

This theorem was proved by Georg Pick in 1899. It was first published in 1899, but only gained attention in 1969 through the popular *Mathematical Snapshots* by H. Steinhaus.[8] The theorem gives an elegant formula for the area of simple polygons. The derived formula does not require math proficiency beyond elementary school and can be easily verified with the help of a geo board. One of its applications is in forestry. It was used to calculate the area inside a polygon area drawn on a map. A special machine was used to first graph the dots on the maps and then the theorem was applied.

LEARNING OBJECTIVES

Students will be able to find the area of simple polygons using Pick's theorem, which is a good introductory geometry topic that considers polygons with vertices that have integer coordinates.

DEFINITIONS

Area. The quantitative measure, in square units of the interior surface of a two-dimensional figure.

Grid. A patterns of horizontal and vertical lines, usually forming squares.

First the students will need to understand the grid. A grid is a set of points in the plane that are a distance of 1 from each other. This distance is considered only horizontally and vertically. The following diagram shows a visual way of imagining a grid.

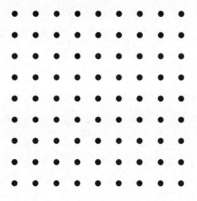

A polygon is a figure that has three or more vertices connected by line segments. This can be created on the geo board in numerous ways, and when the vertices are on the grid points, it is a simple polygon.

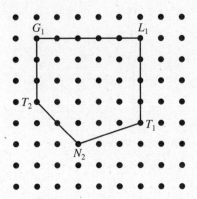

Pick's theorem: Let P be given simple polygon. Then its area is equal to $K = I + (b/2) - 1$, where I equals the number of inside grid points, and b is the number of grid points on the sides of the polygon, and K is the area of polygon P.

Example 5

Using Pick's theorem, calculate the area of the polygon in the following figure.

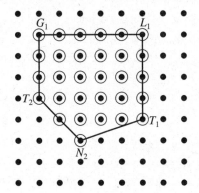

Solution
The number of interior points is equal to 15. The number of points on the sides is equal to 15, and the area is equal to $K = 15 + (15/2) - 1 = 22.5$.

ADDITIONAL PRACTICE

Using a geo board, students create different polygons and compare how the interior points affect the overall area. Students can also consider the perimeter of the shapes as the interior points increase or decrease.

For additional practice: Go to the website http://www.cs.drexel.edu/~crorres/Archimedes/Stomachion/Pick.html. This is an interactive geo board to reinforce this concept.

PROBLEMS

Build each shape on your geo board and find the area of them.

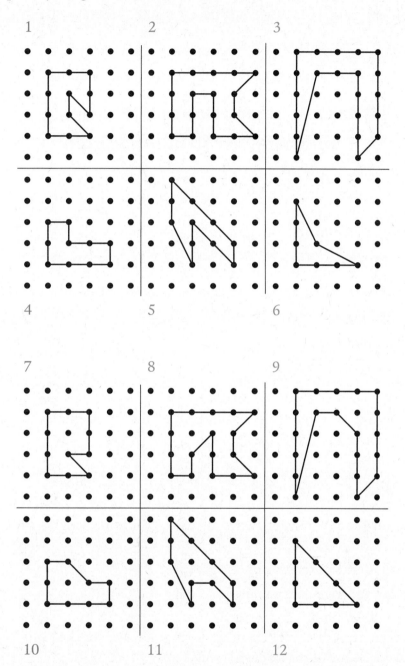

13. Suppose that the vertices of a polygon all lie on a rectangular lattice of points where adjacent points on the lattice are a distance 1 apart. Then the area of the polygon can be found using Pick's formula: $K = I + b/2 - 1$, where I is the number of lattice points inside the polygon and b is the number of lattice points on the boundary of the polygon. Pat applied Pick's formula to find the area of a polygon but mistakenly interchanged the values of I and b. As a result, Pat's calculation of the area was too small by 35. Using the correct values for I and b, the ratio $n = I/b$ is an integer. Find the greatest possible value of n.

SOLUTIONS

1. 5	6. 7	11. 5
2. 8	7. 5.5	12. 4.5
3. 10.5	8. 8.5	13. 15
4. 4	9. 10	
5. 4.5	10. 4.5	

Equilateral versus Equiangular

OVERVIEW

"Equilateral" and "equiangular" are two special types of polygons with some interesting qualities. A polygon is equilateral if all its sides are the same length, like a square or a regular five-pointed star. A polygon is equiangular if all its angles have the same measure; not all equiangular polygons are equilateral, and not all equilateral ones are equiangular. Equilateral triangles in particular have lots of useful properties; this unit covers Viviani's theorem as an example. You can learn more about Viviani's theorem in Chapter 17.

LEARNING OBJECTIVES

Students will be able to describe the attributes of equilateral and equiangular polygons. Further, they will see an interesting application of the mathematical relations between angles and sides of a figure.

DEFINITIONS

Regular polygon. A polygon with all sides the same length and all angles the same measure.

Equilateral. All sides are the same length.

Equiangular. All angles are equal.

Congruent. Relating to geometric figures that have the same size and shape. Two triangles are congruent; for example, if their sides are of the same length and their internal angles are of the same measure.

Theorem. A proved mathematical generalization.

Invariant. Unvarying; constant.

Viviani's theorem. States that the sum of the distances from any interior point to the sides of an equilateral triangle equals the length of the triangle's altitude.

A polygon is called equilateral if all of its sides are equal. Common examples of equilateral polygons are rhombi, squares, and regular polygons such as equilateral triangles.

A polygon is equiangular if all of its angles are congruent.

The only equiangular triangle is the equilateral triangle. Rectangles, including the squares, are the only equiangular four-sided figures.

For an equiangular n-gon, each angle is $180° - (360°/n)$; this is the equiangular polygon theorem.

Pentagon: $180° - (360°/n) = 180° - (360°/5) = 180° - 72° = 108° \times 5 = 540°$.

Students can create equilateral triangles and use a protractor to prove the angles add to $180°$ and prove or disprove the equality of angles. Students can create other n-gons and measure their angles to prove the equiangular n-gon theorem.

PROBLEMS

1. Prove the equiangular polygons theorem.
2. Each angle of a polygon is equal to $160°$. How many sides does the polygon have?
3. Each angle of a polygon is equal to k, where k is a whole number. How many different values are possible for k?
4. Construct an equiangular hexagon whose side lengths are 1, 2, 3, 4, 5, and 6, in some order.

   (*Hint:* Start with an equiangular triangle of side 9 and cut corners!)
5. The side lengths of an equiangular hexagon are labeled, consecutively, $a_1, a_2, a_3, a_4, a_5, a_6$. Prove that $a_1 - a_4 = a_5 - a_2 = a_3 - a_6$.
6. Prove Viviani's theorem for equilateral polygons.
7. Prove Viviani's theorem for equiangular polygons.

SOLUTIONS

1. The sum of the internal angles of an n-gon is $(n-2) \times 180° = n \times 180° - 360°$. Because all angles of an equiangular polygon are equal, each angle is equal to $(n \times 180° - 360°)/n = 180° - (360°/n)$.
2. The equation $180° - (360°/n) = 160°$ can be written as $180° - 160° = 360°/n$, which implies $20° = 360°/n$, yielding $n = 18°$.
3. Note that $k = 180° - (360°/n)$ is a positive integer. Hence, n is a divisor of 360 that is greater than 2. Because $360 = 2^3 \times 3^2 \times 5$, it has $(3+1)(2+1)(1+1) = 24$ divisors, including 1 and 2, the answer is 22.
4. All angles of the hexagons with side lengths 1, 5, 3, 4, 2, 6 are $120°$.

5. Same strategy as in the previous problem.
6. The sum of the areas of the *n* triangles whose bases are the sides of the polygon is equal to the area of the polygon.
7. Consider a regular polygon such that the given polygon lies in its interior and use the idea in the previous problem.

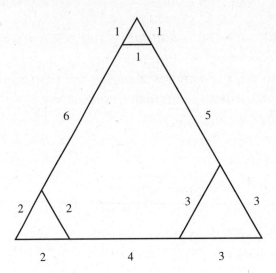

Math and Chess

OVERVIEW

Mathematics and chess have been pursued intellectually for centuries by many researchers and scientists, especially mathematicians, who greatly appreciated the logic and symmetry in chess. The analysis of chess can be extremely complicated due to many possible options at each move.

LEARNING OBJECTIVES

- Recall (or learn) the chess pieces' moves.
- Visualize the 64 squares of the chessboard, its rows, columns, and diagonals.
- Use creative thinking to solve the questions asked.

Have the students familiarize themselves with a chess board. It could be an image on a PowerPoint presentation or it could be an actual chess board. Also present the pieces.

We can easily formulate questions like, how many pieces of a given type can be placed on a chessboard without any two attacking each other?

Example 6

What is the greatest number of bishops we can place on a regular 8 × 8 chess board without any two attacking each other?

Solution

The answer is 14. Because bishops move only diagonally, we can divide the problem in two parts: on the number of bishops on white and black cells. We have one main diagonal of eight white cells and six more white diagonals parallel to it. So, at most, we can place seven bishops that do not attack each other on white squares. Using symmetry reasoning, we can place at most seven bishops that do not attack each other on black squares. So the greatest number is 14. We can place the bishops in the following manner: A1, A2, A3, A4, A5, A6, A7, A8, and H2, H3, H4, H5, H6, H7.

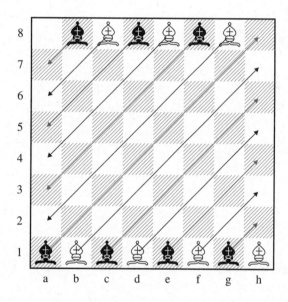

What is the greatest number of knights we can place on a regular 8 × 8 chess board without any two attacking each other?

Answer: 32 (all knights will be on white squares)

Can a knight start in one of the corners and arrive at the opposite corner traveling through each square precisely once?

Answer: No, because it will need 63 moves to arrive at the opposite corner and at each move the knight changes the color of the square it occupies. So after an odd number of moves (63) it will be on a square of different color while the opposite corner squares have the same color.

Area and Volume of a Sphere

OVERVIEW

The area of a sphere of radius r is $A = 4\pi r^2$.
The volume of a sphere of radius r is $V = (4/3)\pi r^3$.

LEARNING OBJECTIVES

- Review the formulas for the area and volume of a sphere.
- Offer an application of these formulas into to a real-life problem.
- Provide an exercise of visualizing a sectional view of a sphere.

PROBLEMS

A box holds eight chocolates (assumed spherical). A front/top/side view of the box is shown. An identical box holds a single chocolate having twice the diameter of the chocolates in the original box. A view of the box is shown:

1. Determine which box contains more chocolate by comparing the volume of chocolate in each box.
2. Each chocolate is wrapped up in paper. Determine which box requires more wrapping paper to wrap the chocolate. Justify your answer.

Solution

We recall that the volume and area of a sphere of radius r are $V = (4/3)\pi r^3$ and $A = 4\pi r^2$. Let the diameter of each chocolate in the smaller box be d.

1. Then the volume of the chocolate in the first box is $V_1 = 8(4/3)\pi(d/2)^3 = (4/3)\pi d^3$, while the volume of the chocolate in the second box is $V_2 = (4/3)\pi d^3$. Hence, the two boxes contain the same amount of chocolate.
2. The total area of the chocolate wrapping paper in the first box is $A_1 = 8 \times 4\pi(d/2)^2 = 12\pi d^2$, while the area of the chocolate wrapping paper in the second box is $A_2 = 4\pi d^2$. Hence, the first box requires more wrapping paper to wrap the chocolate.

Notes

1. Titu Andreescu and Branislav Kisacanin, *Math Leads for Mathletes*, Book 1 (Providence, RI: American Mathematical Society, 2014).
2. Source: Purple Comet Math Meet! Contest 2019.
3. Titu Andreescu and Branislav Kisacanin, *Math Leads for Mathletes*, Book 1 (Providence, RI: American Mathematical Society, 2014).

4. Andreescu and Kisacanin, *Math Leads for Mathletes*, Book 1.
5. Andreescu and Kisacanin, *Math Leads for Mathletes*, Book 1.
6. Source: AwesomeMath Admission Test A, 2019.
7. Source: *Math Leads for Mathletes,* Book 1.
8. Władysław Hugo Dionizy Steinhaus was a Polish mathematician and educator.

SECTION III

Full Units

In this final section, you are provided with full units that range in difficulty from the easiest, "Angles and Triangles" to the hardest, "Nice Numbers." That said, each unit has problems that are accessible along with problems that will provide a significant challenge for high-level students. The provided examples will bring to life the important concepts and takeaways for the lesson. Be sure to utilize the relate, reflect, and revise process so that the lessons can evolve and grow over time.

The chapters have Learning Objectives, Definitions, Vocabulary, Examples, Problems, and Solutions. Some will have interesting facts from history and mathematics when appropriate. All lessons are intended to engage and delight mathematics students of all ages.

For further problems and training resources that can be utilized in problem-based learning classrooms, visit http://awesomemath.org!

CHAPTER 9

Angles and Triangles

Learning Objectives

Students can identify angles and types of triangles. Angles and triangles are fundamental structures in geometry. There are many ways to qualify them, and many important results that follow. Finding relationships between angles is a good exercise for interpreting geometric diagrams and extracting meaningful information. This unit is accessible to middle and high school students.

Definitions

Right angle. An angle measuring 90°.
Perpendicular lines. Two lines that intersect at a 90° angle.
Parallel lines. Two lines that run side by side and never cross or intersect.
Straight angle. An angle measuring 180°.
Supplementary angles. Two angles that added together produce an angle of 180°.
Congruent triangles. Two triangles that are the same size and shape.
Similar triangles. Triangles that have at least two angles that are congruent.

Angles and Parallel Lines

Assume you have a point P in the plane. If you draw a full circle around this point you get 360° rotation. Now consider a line AB that passes through P and P is in between A and B. Then $\angle PAB$ is half of this circle and has 180°. Now imagine a circle that is divided in four equal parts by two lines *AB* and *CD* that pass through *P*. Then these lines divide the full angle of 360° in four equal angles. These angles are equal to 90° and are called right angles. The lines *AB* and *CD* intersecting at a right angle are called perpendicular.

There are some nice properties for parallel lines. Any line intersecting two parallel lines forms with each of parallel lines equal angles. Any pair of angles is supplementary if their measurements add up to 180°.

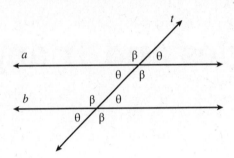

We can see that when two lines (lines *a* and *b*) are intersected by a line called a transversal (line *t*). This produces two sets of congruent angles (β and θ) equal angles and four supplementary angles.

Now consider a triangle *ABC*. We want to prove that the sum of angles in the triangle is 180°. Draw a parallel line though *A* to *BC*. Let *X* and *Y* be two points on that line such that segment *XY* contains point *A*.

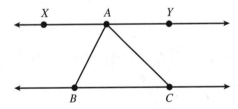

Then $\angle XAB = \angle ABC$ and $\angle YAC = \angle ACB$.

Therefore $\angle ABC + \angle BAC + \angle ACB = \angle XAB + \angle BAC + \angle CAY = 180°$.

Example 1

Consider a polygon with vertices $A_1 A_2 \ldots A_n$. Prove that the sum of angles in *n*-sided polygon is $(n-2) \times 180°$.

Solution
Consider vertex A_n and let us triangulate the polygon into the following triangles: $A_n A_1 A_2$, $A_n A_2 A_3, A_n A_3 A_4, \ldots, A_n A_{n-2} A_{n-1}$.

There are $n-2$ triangles in total. Note that the sum of angles in these triangles is equal to the sum of all angles in the polygon. Thus, the sum of angles in n-sided polygon is equal to $(n-2) \times 180°$.

Note that congruent triangles are similar triangles, but the converse is not true.[1]

Also, because we call these triangles similar, they look the same and one triangle is a rescaling of the other triangles by some constant k. It follows that:

$$\frac{a}{a'} = \frac{b}{b'} = \frac{c}{c'} = k.$$

Example 2

Assume we have two similar triangles ABC and $A'B'C'$ such that line segments $AB = 4$, $BC = 6$, $CA = 8$, and $A'B' = 6$. Find $B'C'$ and $A'C'$.

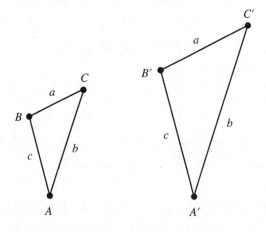

Solution

Because triangles ABC and $A'B'C'$ are similar we have

$$\frac{AB}{A'B'} = \frac{BC}{B'C'} = \frac{CA}{C'A'}$$

$$\frac{4}{6} = \frac{6}{B'C'} = \frac{8}{C'A'}$$

and $B'C' = 9$, $C'A' = 12$.

Example 3

In the following figure, ∠ACB = 40°. If triangles ACD and CBA are similar, find the angle measure of ∠EAC.

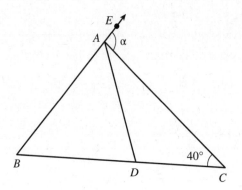

Solution

Since triangles ACD and CBA are similar, we find that ∠ACD = ∠CBA, ∠CAD = ∠BCA, and ∠ADC = ∠CAB. Thus, ∠BAC = 80°. Therefore, α = 80°.

Summary

Using a protractor, draw some similar triangles and test to see if the angles always add up to 180°. Does the sum of the interior angles change with the length of the line segments?

Draw similar triangles (triangles with at least two equal angles). Use your protractor to prove their similarity. Then, answer the following questions:

1. What is the relationship between the line lengths of the similar triangles?
2. How does this affect the interior angle measurement?
3. Can you draw two similar triangles with only two congruent angles?

PROBLEMS

1. Let ABCD be a quadrilateral such that ∠A = 70°, ∠B = 80°, and ∠C = 90°. Find ∠D.
2. Consider an isosceles triangle with one angle equal to 100°. Find the other two angles.
3. Find the sum of angles in a hexagon.
4. On the following picture, lines AX and CY are parallel, ∠XAB = 40° and ∠YCB = 60°. Find ∠ABC.

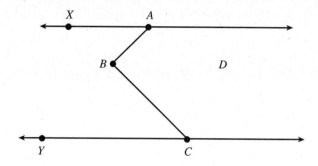

5. In the figure below, lines L_1 and L_2 are parallel. Find the value of θ.

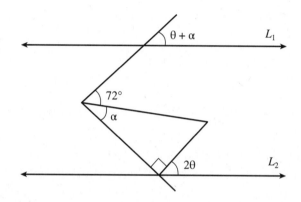

6. One of the angles of a regular n-gon is $135°$. Find the value of n.

7. An n-sided polygon has two interior angles of sizes $94°$ and $51°$. The remaining interior angles are all equal in size and have an integer value. If $4 \leq n \leq 20$, determine the value of n.

8. Find $a+b+c+d+e+f+g$ in the figure below.

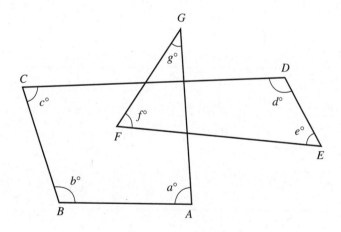

9. In the quadrilateral $ABCD$, E is the midpoint of AB, F is the midpoint of BC, and G is the midpoint of AD. If GE is perpendicular to AB and $\angle ADC = 70°$, find the measure $\angle GCD$.

10. In the figure below, AC = 10, AD = 6, FC = 5, and DF = 3. Find the length of BE.

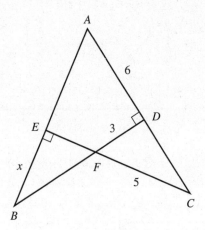

SOLUTIONS

1. The sum of the angles in a quadrilateral is 360°. Thus, $\angle A + \angle B + \angle C + \angle D = 360°$, hence $70 + 80 + 90 + \angle D = 360°$, yielding $\angle D = 120°$.

2. An isosceles triangle has an axis of symmetry. Hence, two of its angles are equal. They cannot both be 100°. Otherwise, the sum of angles will be greater than 180°. Let these angles be equal to x. Thus $2x + 100° = 180°$, so $x = 40°$.

3. Divide the hexagon into four triangles by using the diagonals. Thus, the sum of the angles is $4 \times 180° = 720°$.

4.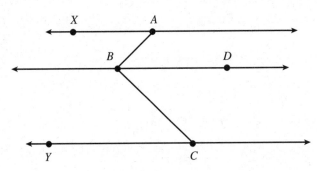

   We draw a parallel line through B and let D be the point as shown in the figure above. Then $\angle XAB = \angle ABD$ and $\angle CBD = \angle DCY$. Thus $\angle ABC = \angle ABD + \angle CBD = \angle XAB + \angle BCY = 40° + 60° = 100°$.

   Remark. In the figure below, lines L_1, L_2 are parallel. Then, $\gamma = \alpha + \beta$.

   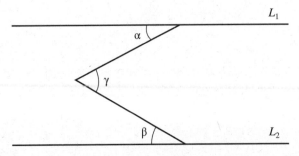

5. It is known that $\theta + \alpha + 90° - 2\theta = 72° + \alpha$. Therefore, $90° - \theta = 72°$. Hence, $\theta = 18°$.

6. The sum of all angles is $n \times 135°$. Also, we know that the sum of the angles in an n-sided polygon is equal to $(n-2) \times 180°$. Hence $n \times 135° = (n-2) \times 180°$ or $n \times 45° = 360°$, yielding $n = 8$.

7. Let α be the size of $(n-2)$ interior angles. We have: $(n-2)180° = (n-2)\alpha + 94° + 51°$. Hence, $(n-2)(180° - \alpha) = 145° = 5° \times 29°$. Because $4 \leq n \leq 20$, we find that $n = 7°$ and $\alpha = 151°$.

8. Let P be the intersection point of lines AG and CD. Let $\angle CPA = \alpha$, $\angle CPG = \angle APD = \beta$. Since, the interior angles of the quadrilateral $ABCP$ sum to $360°$, we have $a + b + c + \alpha = 360°$. Since the sum of the interior angles of the five-side polygon (i.e. pentagon) $DEFGP$ is $540°$, then $d + e + f + g + \alpha + 2\beta = 540°$. We have $a + b + c + d + e + f + g + 2(\alpha + \beta) = 900°$. Therefore, $a + b + c + d + e + f + g = 900° - 2(\alpha + \beta) = 900° - 2(180°) = 540°$.

9.

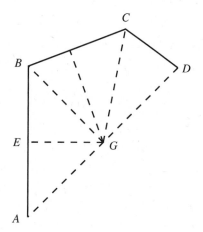

Since GE is both the altitude and median of triangle ABG, this triangle is therefore isosceles. Thus, $AG = GB$. Analogously, triangle GBC is isosceles. Therefore, $BG = GC$. Moreover, since G is the midpoint of AD, we find that $DG = AG = BG = CG$. That is, GCD is isosceles, too. This means that $\angle GCD = \angle GDC = 70°$.

10. Triangles FDC and AEC are similar. Therefore, $FC/AC = DC/EC = FD/AE$. It follows that $5/10 = 4/(5 + EF)$. Therefore $EF = 3$, implying that $EF = FD = 3$. Hence triangles FDC and EFB are congruent, and so $BE = DC = 4$.

Note

1. The proof shows that the sum of angles in n-sided polygon is independent of convexity or concavity of a polygon.

CHAPTER 10

Consecutive Numbers

Learning Objectives

The students will explore different consecutive forms of numbers. They will learn that consecutive forms include even consecutives, odd consecutives, and multiple consecutives, and they might come up with interesting conclusions. The students will learn to represent these numbers algebraically and find patterns in consecutive numbers. Other arithmetic sequences, such as consecutive even or odd integers, have their own emergent qualities. This unit is accessible to middle and high school students.

Definitions

Consecutive. Following one another in uninterrupted succession or order

Integers. . . . −2, −1, 0, 1, 2, . . .

Even numbers. Any integer divisible by 2. The last digit will be 0, 2, 4, 6, or 8.

Odd numbers. Any integer not divisible by 2. The last digit will be 1, 3, 5, 7, or 9.

Numerous problems involve consecutive numbers, such as 1, 2, 3 or 9, 10, 11, 12.
The numbers 15, 18, 21 are consecutive multiples of 3. Many problems involve consecutive even numbers or consecutive odd numbers (for example, 2, 4, 6, 8 or 13, 15, 17). So, how do we represent these numbers algebraically? If the question calls for consecutive numbers, you can let one of these numbers be x. The next consecutive number is $x + 1$, $x + 2$, $x + 3$, etc. (or: $x − 1$, x, $x + 1$). If the question calls for consecutive even numbers, we would have to "make sure" that the number we choose is even. We can do this by letting the first number be $2x$ instead of x. What would the next consecutive number be? Be careful: $2x + 1$ is not even. So, our next number would be $2x + 2$, $2x + 4$, $2x + 6$, etc. Similarly, consecutive odd numbers would be of the form $2x + 1$, $2x + 3$, $2x + 5$, etc.

Example 1

If the sum of two consecutive numbers is 33, what are the numbers?

Solution
Let the first number be x and the second number be $x + 1$. Then: $x + (x + 1) = 33$, which implies that $x = 16$. So our numbers are 16 and 17.

Example 2

The numbers 16, 17, and 18 are consecutive and add up to 51. Find the four consecutive integers that add up to 50.

Solution
The numbers are 11, 12, 13, 14.

Example 3

Write 99 as a sum of consecutive integers in 11 different ways.

Solution
If the integers are $n + 1, n + 2, \ldots, n + k$, then

$$99 = n + n + \ldots + n + (1 + 2 + \ldots + k)$$

$$99 = kn + (k(k+1))/2$$

So, $2 \times 99 = k(2n + k + 1)$.

Then

$k = 2$ and $2n + 2 + 1 = 99$, yielding $n = 48$, or
$k = 3$ and $2n + 3 + 1 = 66$, yielding $n = 31$, or
$k = 6$ and $2n + 6 + 1 = 33$, yielding $n = 13$, or
$k = 9$ and $2n + 9 + 1 = 22$, yielding $n = 6$, or
$k = 11$ and $2n + 11 + 1 = 18$, yielding $n = 6$, or
$k = 18$ and $2n + 18 + 1 = 33$, yielding $n = 3$, or
$k = 22$ and $2n + 22 + 1 = 9$, yielding $n = -7$, or
$k = 33$ and $2n + 33 + 1 = 6$, yielding $n = -14$, or

$k = 66$ and $2n + 66 + 1 = 3$, yielding $n = -32$, or

$k = 99$ and $2n + 99 + 1 = 2$, yielding $n = -49$, or

$k = 198$ and $2n + 198 + 1 = 1$, yielding $n = -99$.

These are all representations, since $2 \times 99 = 2^1 \times 3^2 \times 11$ has $(1+1)(2+1)(1+1) = 12$ divisors different from 1.

Example 4

In how many ways can you get the number 100 by adding some consecutive integers between 1 and 99 inclusive?

Solution

Assuming $a + a + 1 + \ldots + a + k = 100$. Then, $(k+1)a + (k(k+1))/2 = 100$. Thus, $(k+1)(2a+k) = 200$.

Since $2a + k > k + 1$ and $2a + k - (k + 1) = 2a - 1$, we find that $k + 1, 2a + k$ are factors of 200 with different parities. Therefore, since $200 = 8 \times 25 = 40 \times 5$ are the only possible decompositions of 200 in two factors with different parities, we find that, either $2a + k = 25$, $k + 1 = 8$ or $2a + k = 40$, $k + 1 = 5$. In the former case, $a = 9$, $k = 7$, and in the latter case $a = 18$, $k = 4$. That is:

$$100 = 9 + 10 + \ldots + 16 = 18 + 19 + 20 + 21 + 22$$

Thus, we have only two possible ways:

- Have students write algebraic representations of consecutive numbers based on the previous numbers.
- Have students write consecutive even or consecutive odd numbers using algebraic formulas.

PROBLEMS

1. The numbers 21, 22, 23, and 24 are consecutive and add up to 90. Find the five consecutive integers with the same property (their sum is 90).
2. Write 450 as the sum of:
    (a) three consecutive numbers
    (b) four consecutive numbers
    (c) five consecutive numbers
3. What are four consecutive even numbers that have a sum of 92?
4. Find five consecutive numbers that have a sum of zero.

5. Two consecutive odd numbers have a product of 35. What are the numbers?
6. Three consecutive multiples of 5 have a sum of 75. What are the numbers?
7. The product of two consecutive numbers is 12. What are the numbers?
8. The sum of four consecutive integers is 46. What are the numbers?
9. The sum of five consecutive even integers is 50. What are the numbers?
10. From the product of two consecutive numbers if you subtract the sum of the same two numbers, the answer is 5. What are the numbers?
11. Do there exist two consecutive odd numbers with product 45?
12. Do there exist seven consecutive integers with sum of 130?
13. The sum of nine odd consecutive numbers is 2007. Find the greatest of these nine numbers.
14. The sum of 21 consecutive integers is 168. Find the least of these numbers.
15. Let a, b, c, and d be the square of some consecutive positive integers. Prove that $a + b + c + d - 5$ is also a perfect square.
16. Find all n for which there are n consecutive integers whose sum is a prime number.

SOLUTIONS

1. Let the least of them be equal to x. Then $x + (x + 1) + (x + 2) + (x + 3) + (x + 4) = 90$, which implies that $x = 16$. Thus, the numbers are 16, 17, 18, 19, 20.
2.
    (a) 149, 150, 151.
    (b) 111, 112, 113, 114.
    (c) 88, 89, 90, 91, 92.
3. 20, 22, 24, 26.
4. −2, −1, 0, 1, 2.
5. 5, 7.
6. 20, 25, 30.
7. 3, 4.
8. 10, 11, 12, 13.
9. 8, 10, 12, 14.
10. 3, 4.
11. No. Let x be the first integer. Then $x(x + 2) = 45$. Solving for x (by inspection or the quadratic formula), we find no integer solutions.
12. No. Let x be the first integer. Then $x + (x + 1) + \ldots + (x + 6) = 130$. This leads to a noninteger solution for x.

    Alternatively, let y be the fourth integer. Then $(y - 3) + (y - 2) + \ldots + (y + 3) = 130$, or $7y = 130$. Thus, y is not an integer.

13. We are given that

    $$(2x+1)+(2x+3)+(2x+5)+(2x+7)+(2x+9)+(2x+1)+(2x+13)+\\(2x+15)+(2x+17)=2007.$$

    Thus, we find that $(2x + 17) = 231$.

14. Let x be the first integer. Then $168 = x + (x + 1) + \ldots + (x + 20) = 21x + (20 \times 21/2)$. This gives $x = -2$.

    Alternatively, let y be the eleventh integer. Then $168 = (y - 10) + (y - 9) + \ldots + (y + 10) = 21y$. This gives $y = 8$, so the least of the consecutive integers is $8 - 10 = -2$.

15. Assume without loss of generality $a < b < c < d$. Let $a = n^2$, then $b = (n + 1)^2$, $c = (n + 2)^2$, $d = (n + 3)^2$. It follows that:

    $$\begin{aligned}a+b+c+d-5 &= n^2 + (n+1)^2 + (n+2)^2 + (n+3)^2 - 5\\ &= 4n^2 + 2n + 4n + 6n + 1^2 + 2^2 + 3^2 - 5\\ &= 4n^2 + 12n + 9\\ &= (2n+3)^2\end{aligned}$$

16. Let $m, m+1, \ldots, m+n-1$ be n consecutive integers. Their sum is equal to $mn + ((n-1)n)/2 = p$, where p is a prime.

    Clearly, $n = 1$ and $n = 2$ satisfy our condition ($1 + 2 = 3$).

    If $n = 2k$ where $k \geq 2$, then $k(m + k(2k - 1)) = p$. So $k = p$ and $m = 1 + k - 2k^2 = 1 + p - 2p^2$.

    Thus, all integers $n = 2p$ where p is a prime are good.

    If $n = 2k + 1 \geq 1$, then $(2k + 1)(m + k) = p$. So $2k + 1 = p$ and $m = 1 + k = 1 - (p - 1)/2$.

    Thus, all integers $n = p$, where $n = p$ is a prime, are also good.

CHAPTER 11

Factorials!

Learning Objectives

Students will solve problems involving factorials. Factorials are simple functions with many applications in the study of non-negative integers. They tend to be very large numbers, but with simple ratios between them. Factorials represent the number of ways a set of objects can be arranged, so they appear frequently in combinatorics. The manipulation of large numbers using their prime factors is a fundamental exercise in number theory and reveals new ways to think about mathematical operations. This unit is accessible to middle and high school students.

Definitions

Factor(s). One of the two or more whole numbers that are multiplied to get a product – for example, 13 and 4 are both factors of 52.

Factorial. The product of all natural numbers up to and including the given number – for example, $4! = 4 \times 3 \times 2 \times 1 = 24$. The symbol is !.

Wilson prime. A prime number p such that p^2 divides $(p-1)! + 1$.

Factorials grow rapidly. There is no generalized formula for factorials, and we almost always have to calculate them using multiplication.

Compute the first five factorials:

$1! = 1$
$2! = 1 \times 2 = 2$
$3! = 1 \times 2 \times 3 = 6$
$4! = 1 \times 2 \times 3 \times 4 = 24$
$5! = 1 \times 2 \times 3 \times 4 \times 5 = 120$

Example 1

Evaluate $9!/(3! \times 7!)$

Solution

$$\frac{9!}{3! \cdot 7!} = \frac{1 \cdot 2 \cdot 3 \cdot 4 \cdot 5 \cdot 6 \cdot 7 \cdot 8 \cdot 9}{(1 \cdot 2 \cdot 3) \cdot (1 \cdot 2 \cdot 3 \cdot 4 \cdot 5 \cdot 6 \cdot 7)} = \frac{4 \cdot 3}{1} = 12$$

Example 2

Prove that $5! \times 6 \times 7! = 10!$

Solution

$$\begin{aligned}
5! \times 6 \times 7! &= 1 \times 2 \times 3 \times 4 \times 5 \times 6 \times 1 \times 2 \times 3 \times 4 \times 5 \times 6 \times 7 \\
&= (1 \times 2 \times 3 \times 4 \times 5 \times 6) \times (1 \times 2 \times 3 \times 4 \times 5 \times 6 \times 7) \\
&= (1 \times 2 \times 3 \times 4 \times 5 \times 6) \times (7 \times (2 \times 4) \times (3 \times 3) \times (2 \times 5)) \\
&= 1 \times 2 \times 3 \times 4 \times 5 \times 6 \times 7 \times 8 \times 9 \times 10 \\
&= 10!
\end{aligned}$$

Example 3

Evaluate $12!/(6! \times 7!)$

Solution

$$\frac{12!}{6! \cdot 7!} = \frac{1 \cdot 2 \cdot 3 \cdot 4 \cdot 5 \cdot 6 \cdot 7 \cdot 8 \cdot 9 \cdot 10 \cdot 11 \cdot 12}{(1 \cdot 2 \cdot 3 \cdot 4 \cdot 5 \cdot 6) \cdot (1 \cdot 2 \cdot 3 \cdot 4 \cdot 5 \cdot 6 \cdot 7)} = 132$$

Example 4

If $(9! + 11!)/(8!x + 10!) = 9$, find the value of x.

Solution

$9! + 11! = 9!(1 + (11 \times 10)) = 9!(111)$. Further, $8!x + 10! = 8!(x + (10 \times 9)) = 8!(90 + x)$. Thus, $(9! + 11!)/(8!x + 10!) = 9!(111)/8!(90 + x) = (9 \times 111)/(90 + x) = 9$. Hence, $111 = 90 + x$, that is, $x = 21$.

Example 5

Simplify the following expressions:

1. $\dfrac{(2n-1)! \cdot 2 \cdot n!}{(2n)! \cdot (n-1)!}$

2. $\dfrac{\frac{1}{3!} + \frac{1}{4!}}{\frac{1}{2!} - \frac{1}{3!} + \frac{1}{4!}}$

3. $\left(\dfrac{1}{n!} + \dfrac{1}{(n-2)!}\right) \cdot \dfrac{(n+2)!}{n^3+1}$

Solutions

1. Write $2 \times n! = 2n \times (n-1)!$, then $\dfrac{(2n-1)! \cdot 2 \cdot n!}{(2n)! \cdot (n-1)!} = \dfrac{(2n-1)! \cdot 2n \cdot (n-1)!}{(2n)! \cdot (n-1)!} = \dfrac{(2n)! \cdot (n-1)!}{(2n)! \cdot (n-1)!} = 1.$

2. $\dfrac{\frac{1}{3!} + \frac{1}{4!}}{\frac{1}{2!} - \frac{1}{3!} + \frac{1}{4!}} = \dfrac{\frac{4+1}{4!}}{\frac{6-4+1}{4!}} = \dfrac{\frac{5}{4!}}{\frac{3}{4!}} = \dfrac{5}{3}$

3. $\dfrac{1}{n!} + \dfrac{1}{(n-2)!} = \dfrac{1 + n(n-1)}{n!} = \dfrac{n^2 - n + 1}{n!}$. Moreover, $\dfrac{(n+2)!}{n^3+1} = \dfrac{(n+2)!}{(n+1)(n^2-n+1)} = \dfrac{(n+2) \cdot n!}{n^2 - n + 1}$. Therefore, we must simplify the following:

$$\dfrac{n^2 - n + 1}{n!} \cdot \dfrac{(n+2) \cdot n!}{n^2 - n + 1} = n + 2.$$

A **Wilson prime** is a prime number p such that p^2 divides $(p-1)! + 1$. Using a calculator, answer the following questions.

Example 6

Is 5 a Wilson prime?

Solution

Note that $(5-1)! + 1 = 4! + 1 = 24 + 1 = 25$. Because 25 divides $4! + 1$, we get that 5 is a Wilson prime.

Example 7

Is 7 a Wilson prime?

Solution
We have $(7-1)! + 1 = 6! + 1 = 720 + 1 = 721$. When we divide 721 by 7 we get 103. But then if we try to divide 103 by 7 we do not get an integer number. Thus, 7 is not a Wilson prime.

Example 8

We have the following statement: $n! + 1$ is a perfect square of some integer. Study the truth of this statement for all positive integers less than or equal to 7.

Solution
We see that for $n = 1,2,3$ the statement is not true. Also, $4! + 1 = 5^2$ and $5! + 1 = 121 = 11^2$, so the statement is true for $n = 4,5$.

If $n = 6$ we have $6! + 1 = 721$ and $26^2 = 676 < 721 < 27^2 = 729$, false.

If we calculate $7!$, we get $7! + 1 = 5040 + 1 = 5041 = 71^2$, so the statement is true in this case.

Students can write the first five factorials and look for patterns in how the answers build. They can also find at which point every factorial ends in zeros.

Remark Finding the positive integer solutions to the equation $m! + 1 = n^2$ is still an open problem (it is indeed the Brocard-Ramanujan problem). We suggest the teacher talk about this to their students, telling them there are so many open problems in the realm of mathematics.

PROBLEMS

1. Which number is greater, $7!$ or $(1 + 2 + 3 + \ldots + 100)$?
2. In how many zeros does $25!$ end?
3. Find the last nonzero digit of $30!$. (For example, the last nonzero digit of $5! = 120$ is 2.)
4. Find the last digit of $(1! + 2! + \ldots + 100!)$.
5. What is the sum of last two digits of $(1! + 2! + \ldots + 2019!)$?
6. Using a calculator, answer the following questions.
    a) Is 11 a Wilson prime?
    b) Is 13 a Wilson prime?

7. Find the greatest n for which $n!$ ends in exactly 33 zeros.

8. It is given that a, b, c, and d are positive integers that satisfy the equation $d! = a! + b! + c!$. Find the largest value of $a+b+c+d$.

9. If n is an integer such that $\dfrac{(n-6)!}{2(n-5)!} = \dfrac{1}{14}$, find.

10. Solve the following equations in positive integers.

    a) $\dfrac{(n+1)! - 2(n!)}{(n-1)!} = 30$

    b) $\dfrac{(n+2)! - (n+1)! - n!}{n! + (n-1)! - (n-2)!} = 42$

    c) $\dfrac{((3n)!)^2}{(3n-1)!(3n+1)!} = \dfrac{21}{22}$

11. If $\dfrac{3^{n!}}{6^{n!}} \div \dfrac{9^{\frac{n!}{2}}}{6^{(n+1)!}} = (6^n)^6$, find the value of n.

SOLUTIONS

1. Calculating 7! we get 7! = 5040. But we remember that if we add $1+2+3+\cdots+100$ (according to simple sums) we get 5050, which is greater than 7!.

2. Note that the number at the end of the number is the number of times we can divide this number by 10. So we need to find how many tens are in the product of 25!. But $10 = 2 \times 5$, hence we need to find how many fives are at the end of this number, because we have many more twos in this product.

   So which numbers are divisible by 5 in 25!? They are 5, 10, 15, 20, and 25. But $25 = 5 \times 5$, so the number of fives in 25! is 6. Thus there are exactly six zeros at the end of 25!.

3. The highest power of 5 that divides 30! is 7. Thus, we must find the last digit of $30!/10^7$. Since $\dfrac{9!}{5 \times 2^7} = 3 \times 3 \times 7 \times 9$ and $\dfrac{10 \times 15 \times 20 \times 25 \times 30}{5^6} = 2 \times 3 \times 4 \times 6$. We can write that:

   $$30!/10^7 = (3 \times 3 \times 7 \times 9)(1 \times 2 \times 3 \times 4 \times 5 \times 6 \times 7 \times 8 \times 9)^2(2 \times 3 \times 4 \times 6)$$

   The remainder of right expression when divided by 10 is 8.

4. Note that 5! is divisible by 10; hence, the digit is 0. Because 10 divides $5! + 6! + \cdots + 100!$, it follows that they have 0 as the last digit. Thus, the last digit of the sum $5! + 6! + \cdots + 100!$ is zero. The last digit of the sum $1! + 2! + 3! + 4!$ is the

last digit of $1+2+6+24$, which is 3. Therefore, the last digit of the whole sum $1! + 2! + \cdots + 100!$ is also 3.

5. One can observe that for all $n \geq 10$, $n!$ ends with at least two zeros. Also,

$$1! + 2! + \cdots + 9! = 1 + 2 + 6 + 24 + 120 + 720 + 5040 + 40320 + 362880 = 409113.$$

Thus, the last two digits are 13, and their sum is 4.

6. a) After calculations we get $10! + 1 = 3\,628\,801$. This number is divisible by 11, but not by 121, so 11 is not a Wilson prime.

   b) We get $12! + 1 = 479\,001\,601$, which is divisible by 169, so 13 is a Wilson prime.

   The only known Wilson primes are 5, 13, and 563.

7. We use the same idea as in problem 4: the number of zeros at the end of the factorial is the number of fives in its product. A simple estimation shows that $125!$ has $25 + 5 + 1$ zeros at the end because there are 25 numbers divisible by 5^2, and one number is divisible by 5^3. So, $125!$ has 31 zeros at the end. It follows that $130!$ has 32 zeros, $135!$ has 33 zeros, and $140!$ has 34 zeros. Thus, the greatest integer for which $n!$ ends in exactly in 33 zeros is $n = 139$.

8. We can assume that $a \leq b \leq c$. Then, $a! + b! + c! > c!$. Thus, $d! > c!$. Therefore, $d \geq 1 + c$. That is,

$$(c+1)! \leq d! = a! + b! + c! \leq 3 \times c!.$$

Hence, $c + 1 \leq 3$. Therefore, $c \leq 2$.

We can check that $a = b = c = 2$ and $d = 3$ is the only solution. Hence, $d + a + b + c = 9$.

9. $\dfrac{(n-6)!}{2(n-5)!} = \dfrac{(n-6)!}{2(n-5)(n-6)!} = \dfrac{1}{2n-10} = \dfrac{1}{14}$. That is, $2n - 10 = 14$, implying $n = 12$.

10. a) $\dfrac{(n+1)! - 2(n!)}{(n-1)!} = \dfrac{(n-1)!(n(n+1)) - 2n(n-1)!}{(n-1)!} = \dfrac{(n-1)!(n^2 + n - 2n)}{(n-1)!} = n^2 - n = 30.$

    That is, $n^2 - n - 30 = (n - 6)(n + 5) = 0$. This implies that $n = 6$.

    b) $\dfrac{(n+2)! - (n+1)! - n!}{n! + (n-1)! - (n-2)!} = \dfrac{n!((n+1)(n+2) - (n+1) - 1)}{(n-2)!(n^2 + n + n - 1 - 1)} = \dfrac{n!(n^2 + 3n + 2 - n - 2)}{(n-2)!(n^2 + 2n)} =$

    $\dfrac{n!(n^2 + 2n)}{(n-2)!(n^2 + 2n)} = n(n-1)$. Hence, $n(n-1) = 42$. The latter equation reduces to

    $(n - 7)(n + 6) = 0$. Therefore, $n = 7$.

    c) Note that:

    $$(3n-1)!(3n+1)! = (3n-1)!$$
    $$(3n-1)!(3n(3n+1)) = (9n^2 + 3n)((3n-1)!)^2$$
    $$((3n)!)^2 = (3n(3n-1)!)^2 = 9n^2((3n-1)!)^2$$

Hence,

$$\frac{((3n)!)^2}{(3n-1)!(3n+1)!} = \frac{9n^2((3n-1)!)^2}{(9n^2+3n)((3n-1)!)^2} = \frac{9n^2}{9n^2+3n} = \frac{3n}{3n+1}.$$

The equation is reduced to $(3n)/(3n+1) = 21/22$; therefore, $n = 7$.

11. Note that $9^{\frac{n!}{2}} = (3^2)^{\frac{n!}{2}} = 3^{n!}$. Therefore,

$$\frac{3^{n!}}{6^{n!}} \div \frac{9^{\frac{n!}{2}}}{6^{(n+1)!}} = \frac{3^{n!}}{6^{n!}} \div \frac{3^{n!}}{6^{(n+1)!}} = \frac{6^{(n+1)!}}{6^{n!}} = 6^{(n+1)!-n!}.$$

Now, we must solve the equation $6^{((n+1)!-n!)} = (6^n)^6 = 6^{6n}$.

Hence, $(n+1)! - n! = 6n$. This implies that $n!n = 6n$; that is, $n! = 6$, therefore, $n = 3$.

CHAPTER 12

Triangular Numbers

Learning Objectives

Students will be able to use triangular numbers to solve problems. Triangular numbers form an interesting sequence. They can be represented with counting or algebra, and many clever relations and identities can be found. This unit introduces students to concepts in algebra from a function that's easy to visualize and is accessible to middle school and high school students.

Definitions

Triangular number. The number of dots in a triangular array.
Integers. The set of whole numbers and their opposites.

Note that the triangular numbers can be justified by the following diagram:

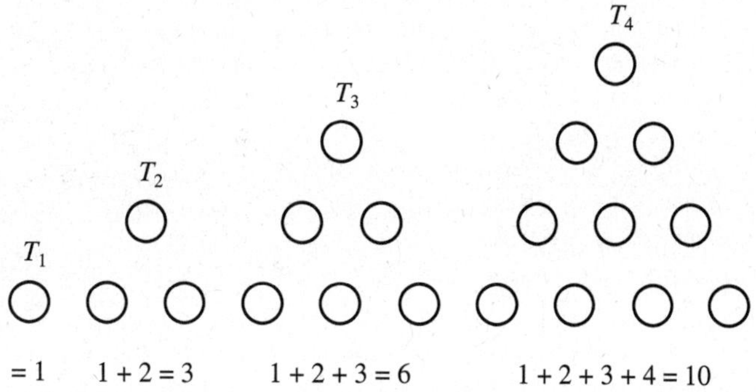

T_1 = 1 T_2: 1 + 2 = 3 T_3: 1 + 2 + 3 = 6 T_4: 1 + 2 + 3 + 4 = 10

$$T_n = (n(n+1))/2$$

A triangular number is the sum of the first n positive integers:

$$T_n = 1 + 2 + 3 + \ldots + (n-1) + n = (n(n+1))/2 = (n^2 + n)/2.$$

The nth triangular number is the number of distinct pairs to be selected from $n + 1$ objects. It solves the "handshake problem" of counting the number of handshakes if each person in a room shakes hands once with every other person. Flocks of birds often fly in this triangular formation.

Even several airplanes when flying together constitute this formation. The properties of such numbers were first studied by ancient Greek mathematicians, particularly the Pythagoreans.[1]

These are the first 100 triangular numbers:

1	3	6	10	15	21	28	36	45	55
66	78	91	105	120	136	153	171	190	210
231	253	276	300	325	351	378	406	435	465
496	528	561	595	630	666	703	741	780	820
861	903	946	990	1035	1081	1128	1176	1225	1275
1326	1378	1431	1485	1540	1596	1653	1711	1770	1830
1891	1953	2016	2080	2145	2211	2278	2346	2415	2485
2556	2628	2701	2775	2850	2926	3003	3081	3160	3240
3321	3403	3486	3570	3655	3741	3828	3916	4005	4095
4186	4278	4371	4465	4560	4656	4753	4950	4950	5050

Example 1

The 36th triangular number is equal to $D + C + L + X + V + I$: the sum of the seven Roman numerals.

Solution

Roman numeral values are as follows: $D = 500$, $C = 100$, $L = 50$, $X = 10$, $V = 5$, $I = 1$. Therefore, the sum of those Roman numerals is equal to the 36th triangular number.

$$500 + 100 + 50 + 10 + 5 + 1 = 666.$$

The 36th triangular number equals 666.

Example 2

The only triangular number that is also a prime is 3. Why?

Solution

No matter what you do, you have two factors $(n(n + 1))/2$.
Students may view the following websites to review simple sums and triangular numbers.
http://www.isallaboutmath.com/triangnum1.aspx (about 5–6 minutes)
http://www.isallaboutmath.com/triangnum2.aspx (about 8–9 minutes)
http://www.isallaboutmath.com/triangnum3.aspx (about 8–9 minutes)

Problems

1. If T is a triangular number, then $9T + 1$ is also a triangular number:

   a) $9T_1 + 1 = 9 \times 1 + 1 = 10 = T_4$

   b) $9T_2 + 1 = 9 \times 3 + 1 = 28 = T_7$

2. Plutarch[2] mentions another method of transforming triangles into squares. He wrote, "Every triangular number taken eight times and then increased by 1 gives a square." His statement is saying that if a triangular number is multiplied by 8, and 1 is added, then the result is a square number.

   If T is a triangular number, then $8T + 1$ is a perfect square:

   a) $8T_1 + 1 = 8 \times 1 + 1 = 9 = 3^2$

   b) $8T_2 + 1 = 8 \times 3 + 1 = 25 = 5^5$

3. At a reception with 25 participants, every two people shake hands with one another. How many handshakes occur?

4. Find the number of diagonals in a dodecagon (12 sides).

Solutions

1.

   a) $9T_1 + 1 = 9 \times 1 + 1 = 10 = T_4$

   Because $T_1 = 1$ and $T_4 + 10$ then

$9(1) + 1$	=	$9 \times 1 + 1$	= 10
10	=	10	= 10

b) $9T_2 + 1 = 9 \times 3 + 1 = 28 = T_7$

Because $T_2 = 3$ and $T_7 + 28$ then

9(3) + 1	=	9 × 3 + 1	= 28
27 + 1	=	27 + 1	= 28
28	=	28	= 28

2. a) $8T_1 + 1 = 8 \times 1 + 1 = 9 = 3^2$

Because $T_1 = 1$, then

8(1) + 1	=	8 × 1 + 1	= 9	= 32
8 + 1	=	8 + 1	= 9	= 9
9	=	9	=9	

b) $8T_2 + 1 = 8 \times 3 + 1 = 25 = 5^5$

Because $T_x = 3$, then

8(3) + 1	=	8 × 3 + 1	= 25	= 5^5
24 + 1	=	24 + 1	= 25	= 25
25	=	25	=25	

3. Each person shakes one another's hands, but it counts double because it occurs separately from each person's point of view. Therefore, (25 × 24)/2 = 300.

4. Since you have a 12-sided figure, you lose the vertex and the two vertices adjacent to it. Therefore, in a 12-sided figure, you have 12 vertices, and you cannot have diagonals from the two adjacent vertices, so you must subtract 3 to include your vertex and the two adjacent vertices that equals 9. Also, you have 6 vertices that are duplicates; therefore, you take the 9 vertices times the 6 duplicates that would equal 54.

PROBLEMS

Directions: Prove these statements to be true.

1. The square of the nth triangular number equals the sum of the cubes of the numbers 1 through n.

2. The sum of the first n triangular numbers is the nth tetrahedral number.

   Tetrahedral formula: $(n(n+1)(n+2))/6$.
3. Calculate the arithmetic mean of the first 20 triangular numbers.

SOLUTIONS

1. Example for the first five triangular numbers

x	y	y²	y³	Σx³
1	1	1	1	
2	3	9	8	
3	6	36	27	225
4	10	100	64	
5	15	225	125	

   x = numbers
   y = triangular numbers
   y^2 = triangular numbers squared
   x^3 = cubes of numbers
   Σx^3 = the sum of cubes of numbers

2. Example for the first four numbers

n	y	Δ#s
1	1	1
2	4	3
3	10	6
4	20	10
Sum of Δ#s		20

   n = number, y = tetrahedral number

3. We know that the nth triangular number is of the form $(n(n+1))/2$. So, we must calculate the following sum:

   $$\frac{1}{20}\left(\frac{1\times 2}{2}+\cdots+\frac{20\times 21}{2}\right).$$

   Note that $(n(n+1))/2 = (n^2 + n)/2$. Therefore, the above sum is equal to:

   $$(1^2 + 2^2 + \ldots + 20^2 + 1 + 2 + \ldots + 20)/40.$$

Since $1^2 + 2^2 + \ldots + 20^2 = (20 \times 21 \times 41)/6$ and $1 + 2 + \ldots + 20 = (20 \times 21)/6$, we find that the above expression is equal to:

$$(20 \times 21 \times 41)/(6 \times 40) + (20 \times 21)/(2 \times 40) = 71.75 + 5.25 = 77.$$

Notes

1. This school of thought was called Pythagoreanism. It originated in the sixth century BC, based on the teachings and beliefs held by Pythagoras and his followers.
2. Lucius Mestrius Plutarchus (c. CE 46–c. CE 120) was a Greek biographer and essayist; he was a contemporary of Nicomachus of Gerasa, an important ancient mathematician best known for his works *Introduction to Arithmetic* and *Manual of Harmonics*.

CHAPTER 13

Polygonal Numbers

Learning Objectives

Students will be able to build polygonal numbers using vertices of polygons and can use polygonal numbers to solve problems. Polygonal numbers are a generalization of triangular numbers, allowing students to create and analyze various sequences. This unit advances the skills from the "Triangular Numbers" unit, Chapter 12, including the construction of sequences and their relations. The unit is accessible to middle and high school students.

Definitions

Polygons. A closed plane figure formed from line segments that meet only at their end points.

Polygonal numbers. A polygonal number is a number represented as dots arranged in the shape of a polygon.

Vertex (Vertices pl.). The point at which two line segments, lines, or rays meet to form an angle.

Polygonal numbers are the number of vertices in a figure formed by a certain polygon. The first number in any group of polygonal numbers is always 1, or a point. The second number is equal to the number of vertices of the polygon. To obtain the third polygonal number, extend by two the sides of the polygon from the second polygonal number and complete the larger polygon by placing vertices and other points where necessary. The third polygonal number counts all vertices and points in the resulting figure:

Triangular Numbers

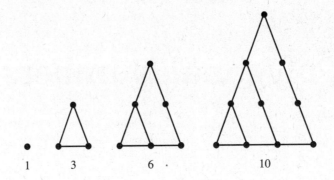

Square Numbers

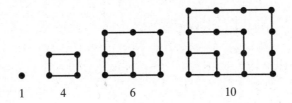

Pentagonal Numbers

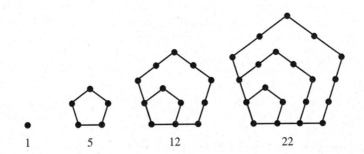

A triangular number is a polygonal number: a number that can be represented by a regular geometric arrangement of equally spaced points. As the name suggests, triangular numbers can be visualized as a triangle of points.

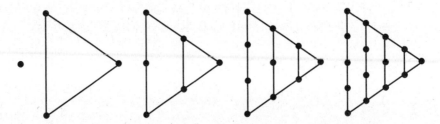

Triangular numbers have a wide variety of relations to other polygonal numbers. Most simply, the sum of two consecutive numbers is a square number. Algebraically,

$$\frac{n^2+n}{2} + \frac{(n-1)^2+(n-1)}{2} = \frac{n^2+n}{2} + \frac{n^2-n}{2} = n^2$$

Alternatively, the same fact can be demonstrated graphically:

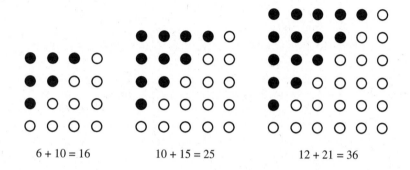

6 + 10 = 16   10 + 15 = 25   12 + 21 = 36

Example 1

Calculate $2T_5 - S_5$ (where is the S_5 fifth square number).

Solution
Since $T_5 = (15)$ and $S_5 = (25)$, $2T_5 - S_5 = 2(15) - (25) = 5$.

Example 2

Calculate $2T_{10} - S_{10}$.

Solution
Since $T_{10} = (55)$ and $S_{10} = (100)$, $2T_{10} - S_{10} = 2(55) - (100) = 10$.

Example 3

Calculate $7!T_7 - 3!S_5$.

Solution
Since $7! = 5040$, $T_7 = 28$, $3! = 6$, and $S_5 = 25$. $7!T_7 - 3!S_5 = 5040(28) + 6(25) = 141\,120 + 150 = 141\,270$.

PROBLEMS

1. Calculate $4T_3 + S_3$.
2. Calculate $2S_4 + T_{12}$.
3. Calculate $12T_{23} - 7S_9$.
4. Calculate $11S_{24} + 2S_5$.
5. Prove that $2T_5 + P_5$ (where P_4 is the fourth pentagonal number) is a perfect square.
6. Prove that $P_6 + T_6 - 2S_6 = 0$.
7. Prove that every other triangular number is a hexagonal number.
8. Prove that $2T_n - S_n = n$.
9. Prove that $2(T_n + P_n)$ is a perfect square.
10. Prove that $P_n + T_n - 2S_n = 0$.
11. Prove that the difference between the nth m-gonal number and the nth $(m+1)$-gonal number is the $(n-1)$-th triangular number. For example, the sixth heptagonal number (81) minus the sixth hexagonal number (66) equals the fifth triangular number (15).

SOLUTIONS

1. Since $T_3 = 6$ and $S_5 = 25$, $4T_3 + S_5 = 4(6) + 25 = 24 + 25 = 49$.
2. Since $S_4 = 16$ and $T_{12} = 78$, $2S_4 + T_{12} = 2(16) + 78 = 32 + 78 = 110$.
3. Since $T_{23} = 276$ and $S_9 = 81$, $12T_{23} - 7S_9 = 12(276) - 7(81) = 3312 - 567 = 2745$.
4. Since $S_{24} = 576$ and $S_5 = 25$, $11S_{24} + 2S_5 = 11(576) + 2(25) = 6336 + 50 = 6386$.
5. Since $T_5 = 15$ and $P_5 = 35$, $2T_5 + 2P_5 = 2(15) + 2(35) = 30 + 70 = 100$, which is the 10th square number.
6. If $P_6 = 51$, $T_6 = 21$, and $S_6 = 36$, then $P_6 + T_6 - 2S_6 = 51 + 21 - 2(36) = 51 + 21 - 72 = 0$. This equation should be balanced, making the statement correct.
7. Every hexagonal number is a triangular number, but not every triangular number is a hexagonal number. Like a triangular number, the digital root in base 10 of a hexagonal number can only be 1, 3, 6, or 9.
8. Two times any triangular number minus the same square number should equal the number that was chosen. Example using the chart below:

N	T_n	S_n
1	1	1
2	3	4
3	6	9
4	10	16
5	15	25

So if we pick the third number, it would be as follows: $2T_3 + S_3 = 2(6) - 9 = 12 - 9 = 3$.
Algebraic proof of Problem 8:

$$2((n^2+n)/2) - n^2 = n$$

$$(n^2+n) - n^2 = n$$

$$n = n$$

9. Two times the sum of a triangular number and a pentagonal number is equal to a perfect square.

Example using the chart below:

N	T_n	S_n	P_n
1	1	1	1
2	3	4	5
3	6	9	12
4	10	16	22
5	15	25	35
6	21	36	51

So if we pick the third number, it would be as follows:
$2(T_n + P_n) = 2(6 + 12) = 2(18) = 36$ (or 6^2, the sixth square number).
Algebraic proof of Problem 9:

$$2\left[\left(\frac{n^2+n}{2}\right) + n^2 + \left[\frac{(n-1)^2 + (n-1)}{2}\right]\right]$$

$$2\left[\frac{n^2+n}{2} + n^2 + \left(\frac{n^2-2n+1+(n-1)}{2}\right)\right]$$

$$2\left[\frac{n^2+n}{2} + n^2 + \left(\frac{n^2-n}{2}\right)\right]$$

$$2\left[\frac{n^2+n}{2} + \frac{n^2-n}{2} + n^2\right]$$

$$2\left[\frac{n^2+n+n^2-n}{2} + n^2\right]$$

$$2\left[\frac{2n^2}{2} + n^2\right] \rightarrow 2\left[n^2 + n^2\right] = 2(2n^2) = 4(n^2)$$

10. So a pentagonal number plus a triangular number minus two times a square number is zero.

    Example using the chart below:

N	T_n	S_n	P_n
1	1	1	1
2	3	4	5
3	6	9	12
4	10	16	22
5	15	25	35
6	21	36	51

    So if we pick the fifth number, you would solve it the following way:

    $$P_n + T_n - 2S_n = 35 + 15 - 2(25) = 50 - 50 = 0.$$

    Algebraic proof:

    $$n^2 + \left(\frac{n^2 - n}{2}\right) + \left(\frac{n^2 + n}{2}\right) - 2n^2$$

    $$n^2 + \left(\frac{n^2 - n + n^2 + n}{2}\right) - 2n^2$$

    $$n^2 + \left(\frac{2n^2}{2}\right) - 2n^2 = 0$$

    $$n^2 + n^2 - 2n^2 = 0$$

    $$2n^2 - 2n^2 = 0$$

    $$0 = 0$$

11. Using the table below to prove the example:

N	T_n	S_n	P_n	HX_n	HP_n
1	1	1	1	1	1
2	3	4	5	6	7
3	6	9	12	15	18
4	10	16	22	28	34
5	15	25	35	45	55
6	21	36	51	66	81

$HP_n - HX_n = T_{n-1}$
81 − 66 = 15

You can do the same for any number in the list.

$HP_n - P_n = T_{n-1}$
or $P_n - S_n = T_{n-1}$

Just replace the variables with the appropriate number.

CHAPTER 14

Pythagorean Theorem Revisited

From the book in preparation *Math Leads for Mathletes – A Rich Resource for Young Math Enthusiasts, Parents, Teachers, and Mentors* – Book 3, by Titu Andreescu and Branislav Kisacanin.

In this unit, we prove the famous Pythagorean theorem in several ways. The Pythagorean theorem is a central topic in early geometry courses. The many proofs of the theorem reflect the connection between algebra and geometry, and the wide variety of problems in this unit can teach students to explore all possibilities of a single theorem.

Rectangular boxes and Euler bricks provide a case study in geometry and number theory. Students can learn about Pythagorean triples and their applications; properties and relations of integers; and the comprehension of three-dimensional geometry. This unit is accessible to middle and high school students.

Learning Objectives

- Prove the Pythagorean theorem and apply it in order to find the length of the solid diagonal of a rectangular box.
- Visualize the face and main diagonals of a rectangular box.
- Apply the Pythagorean theorem to draw a good tridimensional picture that will enable students to correctly visualize the above.

Definitions

Pythagorean theorem. For a triangle with a right angle, the square of the hypotenuse, c, is equal to the sum of the squares of the other two side: $a^2 + b^2 = c^2$.

Euler brick. A rectangular cuboid with integer side dimensions such that the face diagonals are integers.

Pythagorean Theorem

A square drawn on the hypotenuse of a right triangle is equal in area to the sum of the squares drawn on the other two sides of the triangle.

Referring to the following figure, Pythagorean theorem states that $a^2 + b^2 = c^2$.

There are many ways to prove this important theorem. Probably the simplest is to use the following diagram and to notice that the combined area of two small squares in the left, $a^2 + b^2$, is equal to the area of the square c^2 in the right part of the diagram. This is because the rest of the big square is covered by four congruent triangles, arranged in a different way.

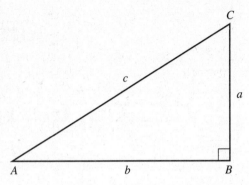

There is another proof based on the right-hand side of the figure. The area of the big square in the figure is $(a + b)^2$. Another way to compute this area is to add the areas of the four right triangles to the area of the square with side c. Thus, $(a + b)^2 = c^2 + (4ab)/2$.

After writing $(a + b)^2$ as $(a + b)^2 = a^2 + 2ab + b^2$, we obtain $a^2 + b^2 = c^2$.

Yet another way to prove Pythagorean theorem is based on the leg theorem. Consider similar triangles obtained by drawing an altitude on the right triangles' hypotenuse as in the following figure.

From $a/m = c/a$, we find $a^2 = cm$. Analogously, from $b/n = c/b$, we get $b^2 = cn$. Finally, when we add the two leg equations, we get $a^2 + b^2 = cm + cn = c(m + n) = c^2$.

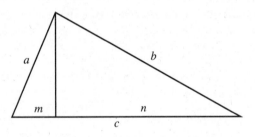

Rectangular Boxes

Example 1

What is the length of a solid diagonal of a $20 \times 30 \times 60$ rectangular box?

Solution

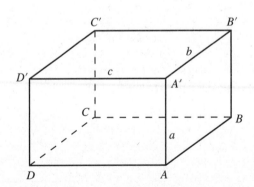

We will solve a slightly more general problem, replacing 20 by a, 30 by b, and 60 by c. $AA' = a$, $A'B' = b$, $A'D' = c$. Also, we will let the face diagonals be $AB' = d$, $A'D' = e$, $B'D' = f$. We need to find the space diagonals length $AC' = g$ in terms of a, b, and c.

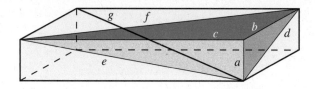

All six faces of the rectangular box are rectangles: $ABCD \equiv A'B'C'D'$, $AA'B'B \equiv DD'C'C$, $AA'D'D \equiv BB'C'C$.

In a rectangle, diagonals are congruent; for example, $AC \equiv BD \equiv A'C' \equiv B'D' = f$. Clearly, the diagonal sections $ACC'A'$ and $BDD'B'$ are also rectangles. The rectangle $ACC'A'$ is pictured here:

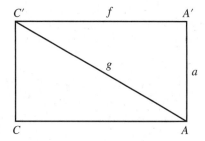

By the Pythagorean theorem, $g^2 = a^2 + f^2$. On the other hand, from the top rectangle $A'B'C'D'$, by the same theorem, $f^2 = b^2 + c^2$. It follows that $g^2 = a^2 + b^2 + c^2$, implying $g = \sqrt{a^2 + b^2 + c^2}$.

In our case ($a = 20$, $b = 30$, and $c = 60$), $g = \sqrt{400 + 900 + 3600} = \sqrt{4900} = 70$.

Example 2

Jamie, a smart sixth grader, has several identical solid bricks. She would like to know the length of the space diagonal of a brick. Jamie has not learned the Pythagorean theorem yet and does not know the formula above. How can she measure the length of this diagonal with a ruler?

Solution
She places three of the bricks like that

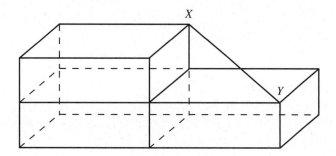

and measures the distance between X and Y with a ruler.

Euler Bricks

A *Euler brick*, named after Leonhard Euler, is a rectangular box in which edges and face diagonals have integer lengths. A primitive Euler brick is an Euler brick whose edge lengths are relatively prime.

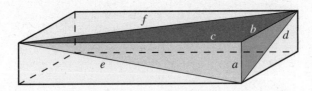

Euler brick with edges a,b,c and face diagonals d,e,f.

The definition of an Euler brick in geometric terms is equivalent to a solution in integers to the following system of equations:

$$\begin{cases} a^2 + b^2 = d^2 \\ a^2 + c^2 = e^2 \\ b^2 + c^2 = f^2 \end{cases}$$

where a,b,c are the edges and d,e,f are the diagonals.

The smallest Euler brick, discovered by Paul Halcke in 1719, has edges (a,b,c) = (44,117,240) and face diagonals (d,e,f) = (125,244,267).

Some other small primitive solutions, given as edges (a,b,c) – face diagonals (d,e,f) are:

(85,132,720) – (157,725,732)
(140,480,693) – (500,707,843)
(160,231,792) – (281,808,825)
(187,1020,1584) – (1037,1595,1884)
(195,748,6336) – (773,6339,6380)
(240,252,275) – (348,365,373)
(429,880,2340) – (979,2379,2500)
(495,4888,8160) – (4913,8175,9512)

A *perfect box* is an Euler brick whose space diagonal also has integer length. In other words, the following equation is added to the system of equations defining an Euler brick: $a^2 + b^2 + c^2 = g^2$, where g is the space diagonal. As of today, no example of a perfect box has been found and no one has proven that none exists.

PROBLEMS

1. On a map, Xville is 20 miles due south of Yton and 21 miles due west from Zfield. How many miles is Zfield from Yton?

2. Determine the perimeter of a rhombus whose diagonals are 20 and 48.

3. A Pythagorean triangle is a right triangle, all of whose sides have integer lengths. Prove that there are infinitely many Pythagorean triangles, two of whose sides are consecutive integers.

4. Prove that for any positive integers m and n such that $m > n$, the following numbers form a Pythagorean triangle:

$$A = m^2 - n^2, B = 2mn, C = m^2 + n^2.$$

This formula is due to Euclid.

5. Let $ABCD$ be a square with area 4 and let P be the point in its interior such that $PA = PB = PM$, where M is the midpoint of CD. Find the area and the perimeter of pentagon $ADCBP$.

6. Find the distance between two opposite corners of a $2 \times 3 \times 6$ rectangular box.

7. A ladder rests against a wall so that its top is 42 ft from the ground. If the ladder is lowered so that its top is 37.5 ft from the ground, the distance from its bottom to the wall increases by 13.5 ft. How long is the ladder?

8. Determine the area of a triangle whose side lengths are 13, 14, and 15.

9. Show that the height h of a right triangle is equal to the geometric mean of the projections of the legs onto the hypotenuse, m and n, i.e. $h^2 = mn$.

SOLUTIONS

1. From the following picture:

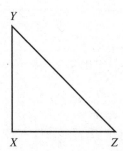

It is obvious that $YZ^2 = XY^2 + XZ^2 = 20^2 + 21^2 = 841 = 29^2$, therefore $YZ = 29$ miles.

2. Diagonals of a rhombus are perpendicular and they bisect each other. From the Pythagorean theorem, we find that the side of the given rhombus is the hypotenuse of a right triangle with sides 10 and 24. Since $10^2 + 24^2 = 100 + 576 = 767 = 26^2$, we find that the perimeter of this rhombus is $4 \times 26 = 104$.

3. Let $a = n$, $b = x$, and $c = x + 1$; then from the Pythagorean theorem, we have

$$n^2 + x^2 = (x+1)^2$$

$$n^2 + x^2 = x^2 + 2x + 1$$

$$n^2 = 2x + 1$$

This tells us that n^2 must be an odd number, so n also must be odd. This is the only limitation, so for any odd length of side a we can find two consecutive integers $b = (a^2 - 1)/2$ and $c = (a^2 + 1)/2$, so that a, b, and c form a Pythagorean triangle. Therefore, there are infinitely many such Pythagorean triangles.

4. We only need to prove that numbers A, B, and C satisfy the Pythagorean theorem. This is indeed true because

$$\begin{aligned} A^2 + B^2 &= (m^2 - n^2)^2 + (2mn)^2 \\ &= m^4 - 2m^2n^2 + n^4 + 4m^2n^2 \\ &= m^4 + 2m^2n^2 + n^4 \\ &= (m^2 + n^2)^2 = C^2 \end{aligned}$$

5. Since the area of $ABCD$ is 4, we know its side is 2. From the following picture

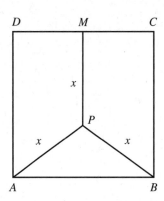

we know that x, $(2 - x)$, and 1 form a right triangle. Therefore

$$(2 - x)2 + 12 = x2$$

$$4 - 4x + x2 + 1 = x2$$

$$4x = 5$$

$$x = 5/4$$

The area $AADCBP = 4 - AABP = (4 - 2(2 - x))/2 = 2 + x = 13.25$.
The perimeter $PADCBP = 6 + 2x = 8.5$.

6. The square of the main diagonal of a rectangular box is $D^2 = a^2 + d^2$, where d is the diagonal on the face of the box with sides b and c. Again, from the Pythagorean theorem, $d^2 = b^2 + c^2$. Finally, $D^2 = a^2 + b^2 + c^2 = 2^2 + 3^2 + 6^2 = 4 + 9 + 36 = 49 = 7^2$; hence $D = 7$.

7. If x is the length of the ladder and y is the distance of its bottom from the wall in the first situation, then the distance from the wall in the second situation is $(y + 13.5)$. We can write the Pythagorean theorem for both situations:

$$x^2 = 42^2 + y^2$$

$$x^2 = 37.5^2 + (y+13.5)^2$$

The second equation can be rewritten as

$$x^2 = 37.5^2 + y^2 + 27y + 13.5^2.$$

When we combine this with the first equation, we get

$$42^2 + y^2 = 37.5^2 + y^2 + 27y + 13.5^2$$

$$\Rightarrow y = (42^2 - 37.5^2 - 13.5^2)/27 = 6.5$$

$$\Rightarrow x = 42.5.$$

8. In general, problems like this one require the use of Heron's formula, which expresses the area of a triangle in terms of side lengths. In this case, however, it suffices to notice that the given triangle can be decomposed into two right triangles: 5 × 12 × 13 and 9 × 12 × 15, when 12 turns out to be the length of the altitude perpendicular to the side whose length is 5 + 9 = 14. Thus, the area is (14 × 12)/2 = 84.

9. From the similarity of triangles, we have $m/h = h/n$, which implies that $h^2 = mn$.

Assessment Problems

1. A Pythagorean triangle is a right triangle all of whose sides have integer lengths. Prove that there is only one Pythagorean triangle whose side lengths are consecutive integers.
2. Chord AB is 6 in. long and tangent to the smaller of two concentric circles. Find the area of the region between the two circles.

CHAPTER 15

Sequences

Learning Objectives

Students will be able to identify and extend arithmetic and geometric sequences. Sequences can take a variety of forms and reflect many topics in mathematics. This unit deals especially with pattern-finding: given the first few terms of a sequence, students must determine the rule which generates the (possibly infinite) sequence. The ability to recognize patterns is an invaluable skill when finding the answer, and thus the solution, to cryptic problems throughout mathematics. Sequences are used every day in fields such as biology and computer science. One of the most famous sequences is the Fibonacci sequence which was created by Fibonacci in order to explain how fast rabbits could breed under certain ideal conditions. This unit is accessible to middle school and high school students.

Definitions

- **Sequences.** An ordered list of objects.
- **Geometric progression.** A sequence of terms in which the ratio between any two successive terms is the same.
- **Arithmetic progression.** A sequence, such as the positive odd integers 1, 3, 5, 7, . . ., in which each term after the first is formed by adding a constant to the preceding term.

This lesson contains the mathematical side of sequences. Let S be a sequence with n terms. The first term of the sequence is denoted by a_1, the second by a_2, the third by a_3, and so on until the last term, which we will denote by a_n. We will call each a_k, $1 \leq k \leq n$, a term of the sequence. In the above definition, we use the letter S to describe an entire sequence. Now that we have introduced the new notation we will refer to a sequence as $(a_n)n \geq 1$. (Note that a_n represents actual terms of the sequence $(a_n)n \geq 1$.)

Example 1

Consider the sequence $(a_n)_{n \geq 1}$, $a_n = (1/n)$. Find a_1, a_2, a_3, a_4, a_5.

Solution

$a_1 = (1/1)$, $a_2 = (1/2)$, $a_3 = (1/3)$, $a_4 = (1/4)$, $a_5 = (1/5)$.

Example 2

Consider the sequence $a_n = 1/(n(n+1))$. Find a_1, a_2, a_3, a_4, a_5.

Solution

$a_1 = (1/1 \times 2) = (1/2)$, $a_2 = (1/2 \times 3) = (1/6)$, $a_3 = (1/3 \times 4) = (1/12)$, $a_4 = (1/4 \times 5) = (1/20)$, and $a_5 = (1/5 \times 6) = (1/30)$.

Introduce a Geometric Progression

A geometric progression is a sequence of the form $a, ar, ar^2, \ldots, ar^n, \ldots$ where the initial term (first term) a and the ratio r are any nonzero real numbers.

Example 3

Let $a = -1$ be the initial term and $r = -1$ the ratio of a geometric sequence. List the first five terms of the sequence.

Solution

$$a_1 = a = -1$$

$$a_2 = (-1) \times (-1) = 1$$

$$a_3 = (-1) \times (-1)^2 = -1$$

$$a_4 = (-1) \times (-1)^3 = 1$$

$$a_5 = (-1) \times (-1)^4 = -1$$

Example 4

Let $a = 2$ be the initial term and $r = 5$ be the ratio of a geometric sequence. List the first six terms of the sequence.

Solution

$$a_1 = a = 2$$
$$a_2 = 2 \times 5 = 10$$
$$a_3 = 2 \times (5)^2 = 50$$
$$a_4 = 2 \times (5)^3 = 150$$
$$a_5 = 2 \times (5)^4 = 1250$$

Note that in a geometric sequence, a term is equal to the previous term times the ratio.
Reminder: An arithmetic progression is a sequence of the form $a, a + d, a + 2d, a + 3d, \ldots, a + nd, \ldots$ where the initial term (the first term) a and the common difference d are any real numbers.

Example 5

Let $a = 1$ be the initial term and let $d = 7$ be the ratio of an arithmetic sequence. List the first five terms of the sequence.

Solution

$$a_1 = a = 1$$
$$a_2 = a_1 + d = a + 7 = 8$$
$$a_3 = a_2 + d = a + 2 \times 7 = 15$$
$$a_4 = a_3 + d = a + 3 \times 7 = 22$$
$$a_5 = a_4 + d = a + 4 \times 7 = 29$$

A common problem in mathematics is to find a formula or general rule for constructing the terms of a sequence. Sometimes only a few terms of a sequence are given and the goal is to identify the sequence. Even though the initial terms of a sequence do not determine that entire sequence (there are infinitely many sequences starting with an initial set of values), knowing the first few terms may help one make an educated guess. Once a guess is made, one can verify it with the given terms.

Example 6

Find formulas for the sequences with the following first five terms:

a) 1, 1/2, 1/4, 1/8, 1/16

b) 1, 3, 5, 7, 9

c) 1, −2, 3, −4, 5

Solution

a) We recognize that the denominators are powers of 2. Hence, the sequence $a_n = 1/2^n$, $n = 0, 1, 2, \ldots$ is a possible match. The given sequence is a geometric progression with initial term $a = 1$ and the ratio $r = 1/2$.

b) We note that each term is obtained by adding 2 to the previous term. The sequence $a_n = 2n + 1$, $n = 0, 1, 2, \ldots$ is a possible match. The given sequence is an arithmetic progression with the initial term $a = 1$ and the common difference $d = 2$.

c) The terms alternate between negative and positive and increase in absolute value. The sequence $a_n = (-1)^{n-1} \times n$, $n = 1, 2, \ldots$ is a possible match.

Sequences exist throughout all academic areas. Students can identify sequences and their impact in any area and write about it.

Students can also write a math sequence of their own, including the formula to determine the next number. They can present these to the classroom.

PROBLEMS

1. 1, 4, 7, 10, 13, ...
2. 1, 2, 4, 8, 16, 32, ...
3. 2, 3, 5, 7, 11, 13, ...
4. 1, 3, 7, 15, 31, 63, ...
5. 2, 5, 14, 41, ...
6. −1, 2, 7, 14, 23, ...
7. 1, 3, 6, 10, 15, 21, ...
8. 1, 1, 2, 3, 5, 8, ...
9. O, T, T, F, F, S, S, ...
10. J, F, M, A, M, J, ...
11. 3, 1, 2, 8, 3, 1, 3, ...

12. 3, 3, 5, 4, 4, 3, 5, 5, ...
13. 2, 4, 6, 30, 32, 34, 36, 40, 42, 44, 46, 50, 52, 54, 56, 60, 62, 64, 66, ...
14. F, 4, E, S, 9, S, E, 5, E, ...
15. E, N, A, O, L, S, L, U, A, ...
16. S, M, H, D, E, H, I, K, M, ...
17. 202, 122, 232, 425, 262, ...
18. C, D, I, L, M, V, ...
19. A, A, A, A, C, C, C, D, F, G, H, I, I, I, I, ...

SOLUTIONS

1. Sequence of numbers that are of the form $3k + 1$. The next number is the sequence is 16.
2. This sequence is of powers of two. The next number in the sequence is 64.
3. This sequence is of prime numbers. The next number in the sequence is 17.
4. The difference of two consecutive terms, are increasing powers of two. The next term in the sequence is 127.
5. Each following terms is 3 times the previous minus 1. The next term in the sequence is 120.
6. First few terms are $12 - 2, 22 - 2, 32 - 2, 42 - 2, 52 - 2$, and the next term in the sequence is $62 - 2 = 34$.
7. The nth term of the sequence is equal to $n(n + 1)/2$. The next term in the sequence is $(7 \times 8)/2 = 28$.
8. This is the Fibonacci sequence (each term is obtained by adding the two previous terms). The next number in the sequence is 13.
9. The letters are the first letters of the numbers (one, two, three, and so on). The next letter is E.
10. The letters are the first letters of the months of the year. The next letter is J.
11. The numbers represent the number of days months of the year (31 days in January, 28 days in February, and so on), one digit at a time. So the next number is 0.
12. The numbers are the number of letters in the names of the numbers (one, two, three, and so on). The next number is 4.
13. None of the numbers shown contains the letter E when spelled out. The next number not to contain the letter E is 2000. (Any number that includes "hundred" or "one thousand" in its name contains an E.)
14. Replace each number by its Roman numeral equivalent and the sequence reads FIVE SIX SEVE, so the next letter is N.

15. The sequence is Paul Sloane spelled backward. So the next letter is P.
16. The letters are the initial of increasing lengths of time: second, minute, hour, day, week, month, quarter, and year. The next letter would be Y.
17. The series of numbers 20 through 28 with their digits grouped in threes. The next number would be 728.
18. The series is an alphabetical list of Roman numerals. So the next one would be X.
19. These are the initial letters of the US states, listed alphabetically (Alabama, Alaska, Arizona, Arkansas, and so on). The next state would be Kansas.

CHAPTER 16

Pigeonhole Principle

Learning Objectives

Students will be able to use the pigeonhole principle to solve interesting problems. The pigeonhold principle is a formalization of an intuitive concept: when fitting too many objects into too few sets, there will be a set containing two or more objects. The problems in this unit can all be rephrased in this way, by demonstrating that the claim in the problem is akin to sorting pigeons into holes. This unit may introduce middle school and high school students to combinatorial thinking and the basics of constructing a valid proof.

Definitions

Co-prime. Integers without any positive common factor other than 1 and −1, or if their greatest common divisor is 1. (Example: 12 and 25 are co-prime. They only share 1 as their common divisor.)

Subset. A set consisting of elements of a given set that can be the same as the given set or smaller.

Intervals. The totality of points on a line between two designated points or end points that may or may not be included. (Example: [1, 2] is an interval including both of its end points, but [1, 2) is an interval that doesn't include 2 as one of its end points; however, it includes the other end point, i.e., 1.)

Integer. Set of whole numbers and their opposites (negatives).

Mean. The sum of the set of numbers divided by the number of elements in the set.

Consecutive. Following one another in uninterrupted succession or order; successive: six consecutive numbers, such as 5, 6, 7, 8, 9, 10.

The pigeonhole principle is one of those mathematical methods (or strategies) that are extremely easy to state and prove yet have highly nontrivial consequences. Its beauty lies in the fact that it justifies "existence" in numerous mathematical contexts without explicitly constructing the desired objects. This principle found its own niche in the vast world of mathematics.

The pigeonhole principle, or Dirichlet's box principle, usually appears in solving problems in algebra, combinatorial set theory, combinatorial geometry and in number theory. In its intuitive form, it can be stated as follows:

If $kn + 1$ objects are distributed among n boxes, one of the boxes will contain at least $k + 1$ objects.

Proof Assume to the contrary that in each box there are at most k objects. Then we would have at most kn objects in total, a contradiction.

The first applications are nice demonstrations that a principle that looks so trivial is actually a subtle tool in proving results. Moreover, the pigeonhole principle provides us with a clear picture of what is going on. An additional object that we added to kn objects is crucial to that picture. The pigeonhole principle enables us to describe the set of $kn + 1$ with some new properties, providing new information.

Example 1

Prove that no matter how we pick six numbers from the set $\{1,2,\ldots 10\}$, there will be two among them, say a and b, such that $a + b = 11$.

Solution
Divide numbers from the set, in pairs (1,10), (2,9), (3,8), (4,7), (5,6). The sum of numbers in each pair is 11. Consider these pairs of numbers as holes and six given pairs as pigeons. By the pigeonhole principle, there will be two numbers from the same pair.

We can find five numbers that do not satisfy the condition: (1,2,3,4,5) or any five numbers.

Example 2

Prove that among any six numbers from the set $\{1,2,\ldots,10\}$, there exist two numbers that are coprime.

Solution
Note that any two consecutive numbers are coprime. So let the holes be five pairs of consecutive numbers: (1,2), (3,4), (5,6), (7,8), (9,10). Our pigeons will be six chosen numbers. By the pigeonhole principle there exist two chosen numbers that are paired. They are coprime and the problem is solved.

Example 3

Prove that among any six numbers from the set $\{1,2,\ldots,10\}$ there is one that is divisible by another one.

Solution
This problem requires a subtle choosing of holes. So, let us divide all the set in the following subsets: (1,2,4,8), (3,6), (5,10), (7), (9).

Note that any two numbers from one of these sets satisfy the condition. That is why we want to choose these sets as holes to apply the principle. We have five holes and six pigeons (our six numbers). By the pigeonhole principle there exist two numbers such that one is divisible by another one. There are combinations of five numbers, for example (4,5,6,7,9), which do not guarantee the existence of such two numbers.

Through the next example, we provide an application of the pigeonhole principle in a branch of mathematical analysis, called Ergodic theory.[1]

Example 4

We choose 5 numbers from the interval (0,1). Prove that there are two numbers a and b ($a < b$) such that $0 < (b - a) < ¼$.

Solution
This example requires a subtle selection of holes. Let us divide the whole interval into the following subintervals: (0,¼),(¼,½),(½,¾),(¾,1). Now, we have 5 pigeons (the numbers) and 4 holes (the intervals). By the pigeonhole principle, there exists two numbers that belong to the same interval. The difference of two numbers in any interval is at most ¼. Thus, our proof is complete.

At the end of this section, we provide two examples concerning combinatorial geometry.

Example 5

We mark 201 points in a 10 × 10 square. Prove that there are three points located in a 1 × 1 square.

Solution
We dissect the square into 100 unit squares. Then, by the pigeonhole principle, there is a square with at least three marked points.

Example 6

Find the largest number of points that can be marked by the vertices of the 20 × 19 table in such a way that no three marked points lie in the vertices of any right-angled triangle.

Solution

We prove in the general case, i.e., $n \times m$ table, the total number is $n + m$. All vertices of the square lie on $n + 1$ horizontal and $m + 1$ vertical lines. Suppose there are at least $n + m + 1$ marked points. Hence, by the pigeonhole principle, at least one horizontal line contains more than one marked point. Hence, at most n marked points are single on their horizontal lines. Analogously, at most m marked points are single on their vertical lines. Therefore, there is a maked point that lies neither single horizontally nor single vertically. Thus, we have a right-angled triangle. Now, by the figure below, we show that $n + m$ is possible. We marked all vertices of squares of the table lying at the left and lower edge of the table except for the lower left corner.

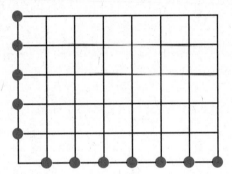

PROBLEMS

1. We have 22 students in a class. Prove we can choose four students who were born the same day of the week.

2. How many times must I throw two dice in order to be sure I get the same score at least twice?

3. A computer has been used for 99 hours over a period of 14 days, a certain number of hours each day. Prove that on some pair of consecutive days, the computer was used at least 15 hours.

4. Twenty-one boys have a total of $200 in notes. Prove that it is possible to find two boys who have the same amount of money.

5. In the following figure, see the paths connecting the square A to the school B. In the square A, there are k students starting to go to the school. They have the possibility to move to the right and up. If they are free to choose any allowed path, find the maximum value of k that in any case there are at least two students who follow the same path.

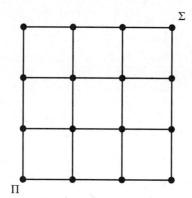

6. Do there exist five distinct prime numbers for which the sum of any three of them is a prime number?

7. We say that two cells of the 10 × 10 table are friendly if they have at least one common vertex. We put a positive integer less than or equal to 10 inside each cell of the table, so that the numbers in friendly cells are relatively prime. Prove that there is at least one number that appears at least 17 times.

8. The teacher asks Valery to choose some of positive divisors of 2009^{10} such that no chosen number divides another. Find the maximum value of divisors that Valery can choose.

9. Find the minimum positive integer n such that among n distinct positive integers, there always exists two distinct positive integers such that either their sum or their difference is a multiple of 2018.

10. Five congruent equilateral triangles could be positioned in a way to cover an equilateral triangle A (these triangles can overlap or run outside the triangle A). Prove that the triangle A could completely be covered by 4 of them.

SOLUTIONS

1. Suppose that there are at most three students born at the same day of a week. Then, the total number of students in the class would at most be $3 \times 7 = 21$. That is impossible.

2. There are six possible scores throwing one die: 1, 2, 3, 4, 5, 6. Thus, there are 11 possible scores throwing two dice: 2, 3, 4, ..., 12. By the pigeonhole principle, we must throw two dice 12 times to ensure that we get the same score twice.

3. Note that $99 = 7 \times 14 + 1$, so we need 7 holes. We group days 1, 2, ..., 14 by the following approach: (1, 2), (3, 4), (5, 6), (7, 8), (9, 10), (11, 12), and (13, 14). Thus, we have 7 holes. Hence, there is at least one hole with at least 15 pigeons. This implies that, on some pair of consecutive days, the computer was used at least 15 hours.

4. Assume to the contrary that all boys have different amounts of money. Thus, the total amount of the money is at least $0+1+2+\ldots+20 = (20/2) \times 21 = 210$ dollars, which is a contradiction. It follows that we can find two boys who have the same amount of money.

5. In the figure below, there are shown all possible paths starting from the square and leading to the school.

   It is clear that in the nodes A_1, A_2, A_3 and B_1, B_2, B_3, there is only one possible path to choose. Counting, we can find that all possible paths are 20. Thus, if we have 21 students, at least two of them would follow the same path. Hence, $k = 21$.

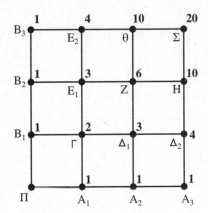

6. Assume that there are such prime numbers. If among them there are three that are not congruent modulo 3 then, by adding each of them to the sum of the other two, we will get three distinct sums modulo 3, therefore, one of them would divisible by 3 and cannot therefore be prime. However, if among the five numbers we only have two distinct remainders when dividing by 3, by the pigeonhole principle there must be three among them that are congruent modulo 3. The sum of these three numbers would be divisible by 3, and therefore not a prime number.

7. Divide the table into 25 squares by side length of 2. In each of these squares, there is at most one even number and at most one number that is divisible by 3. Hence, at most, 50 numbers in the table are divisible by 2 or 3. Now, at least 50 numbers remain, and each of them is equal to 1, 5, and 7. Hence, by the pigeonhole principle, at least one of them appears at least 17 times.

8. Since $2009 = 7^2 \times 41$, all divisors of 2009^{10} would be of the form $7^a \times 41^b$ where $0 \leq a \leq 20$ and $0 \leq b \leq 10$. Thus, we have 11 choices for b. Valery can at most chose 11 divisors. Otherwise, by pigeonhole principle, two of them have the same value of a, and then one certainly divides another. Now, she can chose 11 numbers such as $7^{20-b} \times 41^b$, $b = 0, 1, \ldots, 10$. It is clear that if $7^{20-b} \times 41^b$ divides $7^{20-c} \times 41^c$ for some b,c, then $b \leq c$, $20 - b \leq 20 - c$, implying that $b = c$.

9. Consider the following set of 1010 positive integers: $\{1009, 1010, \ldots, 2018\}$. It is obvious that the sum (or difference) of any two distinct numbers from this set is never a multiple of 2018. This implies that $n \geq 1011$. We shall provide an example for $n = 1011$ as follows.

Since the remainders of a number when divided by 2018 are 0,1,...,2017, then we distribute these remainders into the following 1010 sets:

$$A_0 = \{0\}, A_1 = \{1, 2017\}, A_2 = \{2, 2016\}, \ldots, A_{1008} = \{1008, 1010\}, A_{1009} = \{1009\}$$

If we choose 1011 distinct positive integers, then by the pigeonhole principle, there are two distinct numbers a, b with the remainders belong to the same set, i.e. A_i. If these remainders are different, then $a + b$ is divisible by 2018. If they are the same, then their difference is divisible by 2018.

10. Let us denote the side length of triangle A by a. We mark six points in A, the vertices and the midpoints. Then, there is at least one triangle to pass through two of them. Since in any equilateral triangle, the largest distance between two points is equal to the side length, we find that the side length of each equilateral triangle is greater than or equal to $a/2$. Hence, consider the following triangulation. It is clear that each of the four triangles could be covered by one of the before-mentioned equilateral triangles.

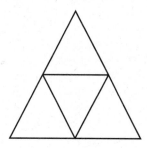

Note

1. Ergodic theory is a branch of mathematics that studies dynamical systems with an invariant measure and related problems. Its initial development was motivated by problems of statistical physics.

CHAPTER 17

Viviani's Theorems

Learning Objectives

In this section, we will study different versions of Viviani's theorem.[1] Viviani's theorem is a geometric fact that's challenging to prove but easy to visualize. The discussion of this theorem incorporates many techniques from geometric proofs: we draw extra lines and structures to support our argument and draw conclusions from geometric observations. This unit is accessible to high school students.

Definition

Viviani's theorem. The sum of distances from a point to the side lines of an equilateral polygon does not depend on the point and an invariant of the polygon.

PROBLEM

If P is an arbitrary point inside of an equilateral triangle, then the sum of the distances from P to the sides of triangle is equal to the length of the altitude of the triangle. That is, this sum remains constant (does not depend on the choice of P).

Solution

We calculate the area of the triangle ABC in two different ways. At first, it is equal to

$$\tfrac{1}{2} \times PD \times BC + \tfrac{1}{2} \times PF \times AB + \tfrac{1}{2} \times PE \times AC.$$

The above expression is equal to $\tfrac{1}{2} \times BC(PD + PF + PE)$. On the other hand, it is equal to $\tfrac{1}{2} \times BC \times h$, where h is the length of the altitude of the triangle. Then, it is easy to deduce that

$PD + PF + PE = h$. This shows that the right-hand side quantity remains constant and does not depend on the choice of P.

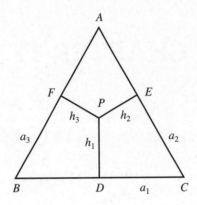

We are ready to study the more general cases. We now distinguish between two types of polygons: equilateral and equiangular.

Theorem (*Viviani's theorem, first form*). In any equilateral convex polygon, the sum of the distances of any interior point to the sides of the polygon is constant.

Proof Connect point P to the vertices of polygon. Let us denote the distances from P to the sides by $d_1, d_2, \ldots$. The area of the polygon is equal to $½d_1 \times a + \ldots + ½d_n \times a$, where a is a side length of the polygon.

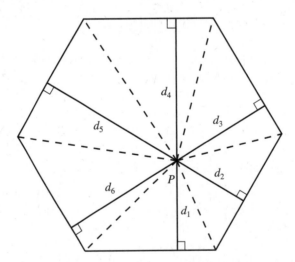

On the other hand, let O be the center of the polygon. Since the polygon is regular, the distance from O to each side is the same. Let us denote such a distance by D. Then the area of polygon is equal to $½D \times a + \ldots + ½D \times a = (n/2)D \times a$.

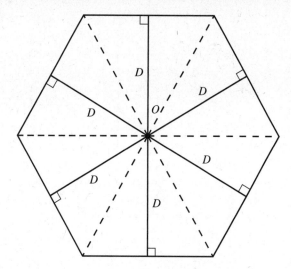

Hence, by comparing the above relations, we find that $d_1 + \ldots + d_n = n \times D$.
This completes our proof.

Remark In the proof above, the distance from O to the sides of regular polygon is called *apothem*.

Theorem (*Viviani theorem, second form*). In any equiangular convex polygon the sum of the distances of any interior point to the sides of the polygon is constant.

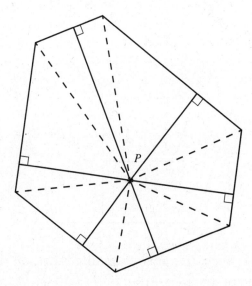

Proof We use the acquired knowledge from Viviani's first theorem. This proof is called an "*embedding proof.*" That is, a regular polygon can be embedded in an equiangular polygon with parallel sides.

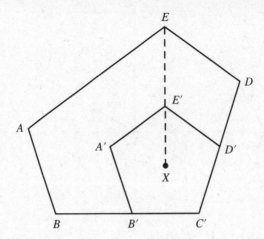

We prove the above claim by considering the case for a pentagon. At first, since all the angles of $ABCDE$ are equal, all have the measure of $108°$. Place the regular pentagon in relation to an equiangular pentagon, as shown. Since $\angle ABC = \angle A'B'C'$, it follows that AB and $A'B'$ are parallel. Analogously, DE and $D'E'$ are parallel. Hence $\angle DEX = \angle D'E'X$ (where the point X lies on the extension of EE'). Moreover, from $\angle DEC = \angle D'E'C'$ we get $\angle AEX = \angle A'E'X$, which means that AE and $A'E'$ are parallel. It follows that a regular pentagon $A'B'C'D'E'$ could always be embedded inside an equiangular pentagon $ABCDE$. So, their corresponding sides are parallel, as shown below.

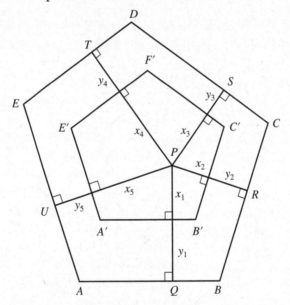

Thus, the distances between the corresponding sides of the two polygons remain constant. Hence, if the sum of distances from point P to the sides of one polygon remains constant, this quantity for the other polygon remains constant as well. According to the first version of Viviani's theorem, this quantity remains constant for regular polygons; hence, it would be constant for equiangular polygons. This completes our proof.

Note

1. Vincenzo Viviani (April 5, 1622–September 22, 1703) was an Italian mathematician and scientist. At the age of 17, he was an assistant of Galileo Galilei.

CHAPTER 18

Dissection Time

Learning Objectives

Students can find the area and perimeter of a geometric figure, dissect it into pieces, and describe the attributes of the resulting shape. Dissections are wordless geometric proofs that can be tackled with clever diagrams. These are difficult problems that show the power of transforming a geometric structure to prove results. These problems can be solved with very little knowledge of geometric formulas, and they can be visualized with physical materials. This unit is accessible to high school students.

Definitions

Dissection. Separating into pieces.
Area. The amount of a surface or plane that a figure covers, expressed in units.
Finite. Having bounds or limits/measurable.
Polygonal surfaces. Area of a polygon.
Transformations. Objects that are rotated, translated, or reflected are transformations of the original object.
Triangulation. To break into triangular shapes using vertices.

The following visual proof of the Pythagorean theorem shows that one can cut any two squares into finitely many pieces and reassemble these pieces to get a square. In fact, much more is true: Any two polygonal surfaces with the same area can be transformed into one another by cutting the first into finitely many pieces and then assembling these pieces into the second polygonal surface.

The properties taught in this lesson were proved by F. Bolyai in 1833 and Gerwien in 1835. It is called the Bolyai-Gerwien theorem.

First, note that using diagonals, a polygon can be cut into finitely many triangles. A triangle can be transformed into a rectangle.

This shows that two squares can be cut and assembled into a single square; thus showing that from a rectangle, a square can be produced.

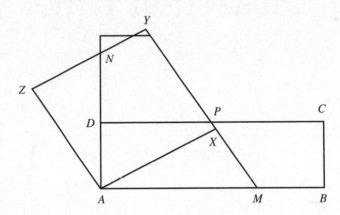

Second, the above proof shows that any polygon can be transformed into a square. It is possible to go backward from a square to a polygon, and therefore transform any polygon into any other polygon of the same area, with a square as an intermediate step.

Here is an example of how we can dissect a triangle into a square.

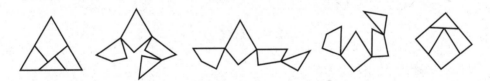

Example 1

Prove that a square can be dissected into n squares for all $n > 6$.

Solution

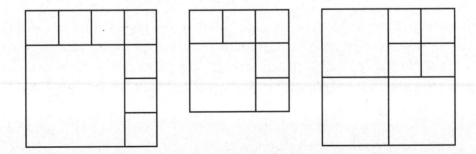

Example 2

Let's triangulate an *n*-gon such that all triangles have vertices of the given *n*-gon. Prove that this triangulation leaves $n - 2$ triangles.

Solution

We can draw one diagonal to dissect an $(n+1)$-gon into an *n*-gon and a triangle. Our inductive hypothesis says that the *n*-gon can be further divided into $n-2$ triangles. Thus, our $(n+1)$-gon gives us $(n-2)+1$ triangles. So $(n+1)$-gon after triangulation will leave us $n - 2 + 1 = n - 1$ triangles.

Example 3

Prove that an equilateral triangle can be triangulated into *n* equilateral triangles.

Solution

An equilateral triangle can be dissected into six, seven, or eight equilateral triangles, as shown in the figure below:

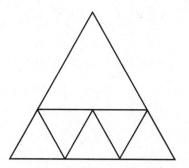 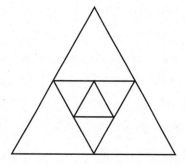

The conclusion follows from an inductive argument. By noting that the triangle can be decomposed into *n* equilateral triangles, then it can be decomposed into $n+3$ triangles by dissecting one of the triangles of the decomposition in four.

Example 4

Dissect a regular octagon in pieces and assemble them to form a square.

Solution

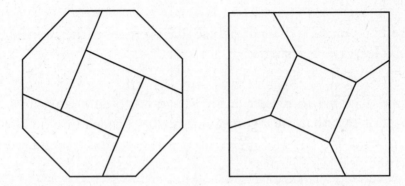

With the whole class, have students demonstrate how they constructed their dissections.

PROBLEMS

1. Dissect a regular hexagon into 6 equal triangles.

2. Dissect a 4 × 9 rectangle into two pieces, which could be assembled to form a square.

3. Take three squares of dimensions 2 × 2, 3 × 3, and 6 × 6. Using only two cuts, can we assemble them into a 7 × 7 square?

4. Assume we have six points in the plane, what is the greatest number of triangles we obtain after triangulating them?

SOLUTIONS

1. Let $A_1A_2A_3A_4A_5A_6$ be our hexagon. Let O be the center of the hexagon. Then, A_1OA_2, A_2OA_3, A_3OA_4, A_4OA_5, and A_5OA_6 are equal triangles.

2.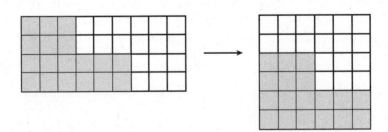

3. There are many ways to accomplish this. One is as follows:
   Label the 2 × 2 unit squares A, the 3 × 3 unit squares B, and the 6 × 6 unit squares C. One of them is the following:

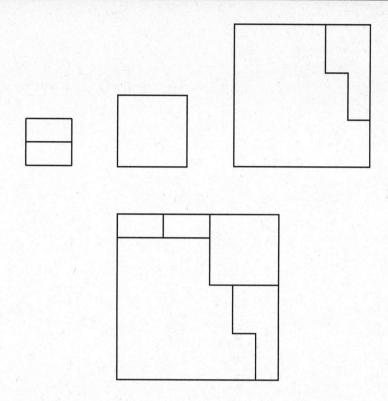

4. We consider all possible distribution of points in the plane. From all of them, triangles with triangles inside, the greatest number of triangles is 7.
   See below:

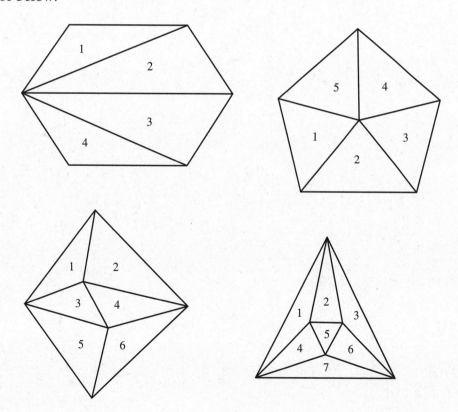

CHAPTER 19

Pascal's Triangle

Learning Objective

Pascal's triangle is a single mathematical structure that is incredibly dense with important subfigures and identities. The patterns in Pascal's triangle relate to structures in combinatorics, geometry, algebra, and number theory, demonstrating the connections between many subjects in this book.

Steps on how to build Pascal's triangle:

- At the top center of your paper write the number "1."
- On the next row write two 1s, forming a triangle.
- On each subsequent row, start and end with 1s and compute each interior term by summing the two numbers directly above it.

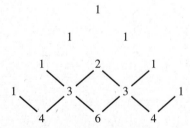

There are some interesting facts in Pascal's triangle:

1. **Important sequences**

   ```
   1
   1   1
   1   2   1
   1   3   3   1
   1   4   6   4   1
   1   5   10  10  5   1
   1   6   15  20  15  6   1
   1   7   21  35  35  21  7   1
   ```

(Left-aligned Pascal's triangle)

The first column contains just one number while the second consists of the positive integers.

The third column contains triangular numbers. Triangular numbers are the numbers of dots it takes to make various-sized triangles.

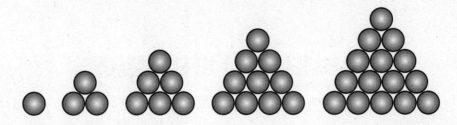

The entries of the fourth column are tetrahedral numbers, or triangular pyramidal numbers. As their name suggests, they represent the number of dots needed to make pyramids with triangular bases.

The columns continue in this way, describing the extrapolations of this triangle/tetrahedron idea to arbitrary dimensions.

2. **Powers of Eleven**

   $1 = 11^0$

   $11 = 11^1$

   $121 = 11^2$

   $1331 = 11^3$

   $14641 = 11^4$

Properties of triangular numbers also indirectly reveal similarities (or patterns) between powers of 11. All we have to do is squish the numbers in each row together. This is easy enough for the first five rows, but what about when we get to double-digit entries? It turns out that we need to carry the tens place over to the number on its left.

$$1, 5, 10, 10, 5, 1 = 1 5^{+1} \, 0^{+1} \, 0 \, 5 \, 1 = 161051$$

Thus, 11^5 is 161 051.

3. Fibonacci Numbers

To obtain the Fibonacci numbers, we sum the entries of left-aligned Pascal triangle:

```
1 1
1 1     1 2
1   2 3   1 5
1   3   3 8   1 13
1   4   6   4 21   1
1   5  10  10   5   1
1   6  15  20  15   6   1
1   7  21  35  35  21   7   1
```

The first eight numbers in Fibonacci sequence are 1, 1, 2, 3, 5, 8, 13, 21, found in Pascal's triangle.

4. Powers of Two

If we sum the entries in each row, we obtain powers of base 2, beginning with $2^0 = 1$.

$$
\begin{array}{c}
1 \\
1 + 1 = 2 \\
1 + 2 + 1 = 4 \\
1 + 3 + 3 + 1 = 8 \\
1 + 4 + 6 + 4 + 1 = 16 \\
1 + 5 + 10 + 10 + 5 + 1 = 32 \\
1 + 6 + 15 + 20 + 15 + 6 + 1 = 64 \\
1 + 7 + 21 + 35 + 35 + 21 + 7 + 1 = 128
\end{array}
$$

Summing the rows reveals powers of 2.

5. Pascal's triangle mod 2.

If we consider Pascal's triangle modulo 2, we find the following figure that is similar to Sierpinski[1] triangle. Moreover, by reading the 32 first rows as the binary expansion of numbers, we get the following sequence: 1,3,5,15,17,51,85,255,257,...

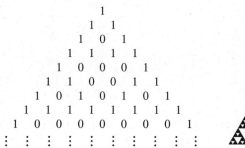
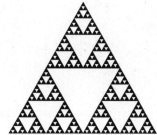

This sequence gives all odd-sided constructible regular polygons. The interesting property of Fermat numbers states that lemniscates (any of several figure-eight or ∞-shaped curves) could be divided into d equal parts with ruler and compass if $d = 2^r p_1, \ldots, p_s$, where r is a nonnegative integer and $p_1, \ldots, p_s$ are distinct Fermat primes.[2] In this figure, you can find the position of Fermat primes in Pascal's triangle.

$$
\begin{array}{ccccccccccccccccc}
 & & & & & & & & 1 & & & & & & & & \\
 & & & & & & & 1 & & 1 & & & & & & & \\
 & & & & & & 1 & & 2 & & 1 & & & & & & \\
 & & & & & 1 & & 3 & & 3 & & 1 & & & & & \\
 & & & & 1 & & 4 & & 6 & & 4 & & 1 & & & & \\
 & & & 1 & & 5 & & 10 & & 10 & & 5 & & 1 & & & \\
 & & 1 & & 6 & & 15 & & 20 & & 15 & & 6 & & 1 & & \\
 & 1 & & 7 & & 21 & & 35 & & 35 & & 21 & & 7 & & 1 & \\
1 & & 8 & & 28 & & 56 & & 70 & & 56 & & 28 & & 8 & & 1 \\
\vdots & & \vdots & & \vdots & & \vdots & & \vdots & & \vdots & & \vdots & & \vdots & & \vdots
\end{array}
$$

Moreover, one can prove the following: If $n \geq 2k \geq 2$, $\binom{n}{k} = 2^{2^m} + 1$, then $k = 1$.

6. Combinations

Recall the combinatorics formula n choose k = $n!/k!(n-k)!$ whenever $k <= n$, and which is zero when $k > n$. We find that in each row of Pascal's triangle, n is the row number and k is the entry in that row, when counting from zero.

$$
\begin{array}{ccccccccccccc}
 & & & & & & \binom{0}{0} & & & & & & \\
 & & & & & \binom{1}{0} & & \binom{1}{1} & & & & & \\
 & & & & \binom{2}{0} & & \binom{2}{1} & & \binom{2}{2} & & & & \\
 & & & \binom{3}{0} & & \binom{3}{1} & & \binom{3}{2} & & \binom{3}{3} & & & \\
 & & \binom{4}{0} & & \binom{4}{1} & & \binom{4}{2} & & \binom{4}{3} & & \binom{4}{4} & & \\
 & \binom{5}{0} & & \binom{5}{1} & & \binom{5}{2} & & \binom{5}{3} & & \binom{5}{4} & & \binom{5}{5} & \\
\binom{6}{0} & & \binom{6}{1} & & \binom{6}{2} & & \binom{6}{3} & & \binom{6}{4} & & \binom{6}{5} & & \binom{6}{6}
\end{array}
$$

The first seven rows of Pascal's triangle are written with combinatorial notation.

So if you want to calculate 6 *choose* 3, look at the 7th row, 4th entry (since we're counting from zero), and you'll find that the answer is 20.

We have the following facts about Pascal's triangle:

1. Sum of entries of nth row of Pascal's triangle is equal to 2^n.
2. The Pascal's triangle has axial symmetry, that is, $\binom{n}{k} = \binom{n}{n-k}$.
3. We can formalize the essence of Pascal's triangle by the following identity:
$$\binom{n}{k} = \binom{n-1}{k-1} + \binom{n-1}{k}.$$

For example,
$$\binom{9}{5} + \binom{9}{6} = \binom{10}{6}.$$

Example 1

If $\binom{11}{5} = \binom{10}{4} + \binom{10}{n}$, find the value of n.

Solution
Note that $\binom{11}{5} = \binom{10}{4} + \binom{10}{5}$. Thus, $\binom{10}{5} = \binom{10}{n}$. That is, $n = 5$.

Example 2

If $\binom{5}{1} + \binom{5}{2} + \binom{6}{3} = \binom{n}{3}$, find the value of n.

Solution
Since $\binom{n-1}{r-1} + \binom{n-1}{r} = \binom{n}{r}$, we deduce that $\binom{5}{1} + \binom{5}{2} = \binom{6}{2}$, $\binom{6}{2} + \binom{6}{3} = \binom{7}{3}$. Hence, $n = 7$.

Summary

We studied some facts about Pascal's triangle: triangular and tetrahedral numbers, Fibonacci sequence in Pascal's triangle, powers of 11 in a Pascal's triangle, powers of two and sum of entries in any row of Pascal's triangle, considering Pascal's triangle modulo 2, and combinatorial aspects of Pascal's triangle. Any of above facts have many interesting aspects. Some of these aspects are presented in following problems.

PROBLEMS

1. Find the smallest value of n such that nth row of Pascal's triangle contains three successive entries with the ratio $3:4:5$.

2. If $\dfrac{1}{\binom{9}{r}} - \dfrac{1}{\binom{10}{r}} = \dfrac{11}{6\binom{11}{r}}$, find the value of r.

3. Taking the product of elements in each row, define the sequence $s_n = \binom{n}{0} \cdot \binom{n}{1} \cdots \binom{n}{n}$. Prove that:

   a.
   $$\dfrac{S_{n+1}}{S_n} = \dfrac{(n+1)^n}{n!},$$

   b.
   $$\dfrac{S_{n+1} \cdot S_{n-1}}{S_n^2} = \left(1 + \dfrac{1}{n}\right)^n.$$

4. Let $\dfrac{1}{2!8!} + \dfrac{1}{3!7!} + \dfrac{1}{4!6!} + \dfrac{1}{(5!)^2} = \dfrac{N}{1!9!}$. Find N.

5. Let's define a set P as following:

$$p = \left\{ (i,j), i,j \in \mathbb{Z} \;\middle|\; \binom{i+j}{i} \equiv 1 \pmod{2} \right\}.$$

   a. Construct the set P around Pascal's triangle for $i,j \in \{0,1,\ldots,30\}$.
   b. Intersect the line $y = (x/4)$ with the set P. Then find the lattice points of intersection.

SOLUTIONS

1. Consider these three successive entries as $\binom{n}{r-1}, \binom{n}{r}, \binom{n}{r+1}$. Then, we can write

$$\dfrac{\binom{n}{r}}{\binom{n}{r-1}} = \dfrac{4}{3}, \quad \dfrac{\binom{n}{r+1}}{\binom{n}{r}} = \dfrac{5}{4}.$$

Moreover, $\dfrac{\binom{n}{r}}{\binom{n}{r-1}} = \dfrac{\frac{n!}{r!(n-r)!}}{\frac{n!}{(r-1)!(n-r+1)!}} = \dfrac{n-r+1}{r}.$

Analogously, $\dfrac{\binom{n}{r+1}}{\binom{n}{r}} = \dfrac{n-r}{r+1}$.

Therefore, $\dfrac{n-r+1}{r} = \dfrac{4}{3}, \dfrac{n-r}{r+1} = \dfrac{5}{4}$.

Hence, $4n = 9r + 3$, $3n = 7r - 3$. Solving the system, we find that $r = 21$, $n = 48$. Thus, the first row with such property is the 48th row.

2.

Note that $\dfrac{1}{\binom{9}{r}} - \dfrac{1}{\binom{10}{r}} = \dfrac{\binom{10}{r} - \binom{9}{r}}{\binom{9}{r}\binom{10}{r}}$.

Moreover, $\binom{10}{r} - \binom{9}{r} = \binom{9}{r-1}$.

Further, $\binom{9}{r} = \dfrac{10-r}{r} \cdot \binom{9}{r-1}$.

Thus, $\dfrac{\binom{10}{r} - \binom{9}{r}}{\binom{9}{r}\binom{10}{r}} = \dfrac{\binom{9}{r-1}}{\binom{9}{r}\binom{10}{r}} = \dfrac{r}{10-r} \cdot \dfrac{1}{\binom{10}{r}}$.

Further, $\dfrac{11}{6\binom{11}{r}} = \dfrac{11}{\dfrac{6 \times 11}{11-r}\binom{10}{r}} = \dfrac{11-r}{6\binom{10}{r}}$.

That is, $\dfrac{r}{10-r} \cdot \dfrac{1}{\binom{10}{r}} = \dfrac{11-r}{6\binom{10}{r}}$.

Hence, $r/(10-r) = (11-r)/6$. Therefore, $r^2 - 27r + 110 = (r-5)(r-22) = 0$. Because $r \leq 11$, we find that $r = 5$.

3.

a. Note that $s_n = \frac{(n!)^{n+1}}{(1!...n!)^2}$. Therefore, $\frac{s_{n+1}}{s_n} = \frac{((n+1)!)^{n+2}(1!...(n+1)!)^{-2}}{n!^{n+1}(1!...n!)^{-2}}$

$= \frac{(n+1)^{n+2} \cdot (n+1)^{-2}}{n!} = \frac{(n+1)^n}{n!}$.

b. Writing $\frac{s_{n+1} \cdot s_{n-1}}{s_n^2} = \frac{\frac{s_{n+1}}{s_n}}{\frac{s_n}{s_{n-1}}}$. By part a, we find that $\frac{s_n}{s_{n-1}} = \frac{(n)^{n-1}}{(n-1)!}$. Hence,

$\frac{s_{n+1} \cdot s_{n-1}}{s_n^2} = \frac{\frac{(n+1)^n}{n!}}{\frac{(n)^{n-1}}{(n-1)!}} = \frac{(n+1)^n}{n^n} = \left(1 + \frac{1}{n}\right)^n$.

4. Multiplying both sides by $10!$, we find that $\frac{10!}{2!8!} + \frac{10!}{3!7!} + \frac{10!}{4!6!} + \frac{10!}{(5!)^2} = N \cdot \frac{10!}{1!9!}$. Therefore,

$$\binom{10}{2} + \binom{10}{3} + \binom{10}{4} + \binom{10}{5} = N\binom{10}{1}.$$

Thus, $\binom{10}{1} + \binom{10}{2} + \binom{10}{3} + \binom{10}{4} + \binom{10}{5} = (N+1)\binom{10}{1} = 10(N+1)$.

Now, we must calculate the sum of the elements of the 10th row in the Pascal triangle. Note that

$$\binom{10}{1} + \binom{10}{2} + \binom{10}{3} + \binom{10}{4} = 10(N+1) - \binom{10}{5} = 10(N+1) - 252.$$

Therefore, $\binom{10}{0} + \binom{10}{1} + \binom{10}{2} + \binom{10}{3} + \binom{10}{4} = 10(N+1) - 251$.

Note that the Pascal triangle has axial symmetry; then,

$$\binom{10}{0} + \binom{10}{1} + \binom{10}{2} + \binom{10}{3} + \binom{10}{4} = \binom{10}{10} + \binom{10}{9} + \binom{10}{8} + \binom{10}{7} + \binom{10}{6}.$$

That is, $2(10(N+1) - 251) = \binom{10}{0} + \binom{10}{1} + \binom{10}{2} + \binom{10}{3} + \binom{10}{4} + \binom{10}{6} + \binom{10}{7} + \binom{10}{8} + \binom{10}{9} + \binom{10}{10}$.

We know that sum of entries of *nth* row of Pascal triangle is 2^n. Therefore,

$$\binom{10}{0}+\binom{10}{1}+\binom{10}{2}+\binom{10}{3}+\binom{10}{4}+\binom{10}{6}+\binom{10}{7}+\binom{10}{8}+\binom{10}{9}+\binom{10}{10}=2^{10}-\binom{10}{5}=2^{10}-252.$$

This implies that $20N + 20 - 2 \times 251 = 2^{10} - 252$.

Therefore, $20N + 20 = 2^{10} + 250$, thus, $N = (1024 + 250 - 20)/20 = 62.7$.

5. The answer is designed in the following figure.

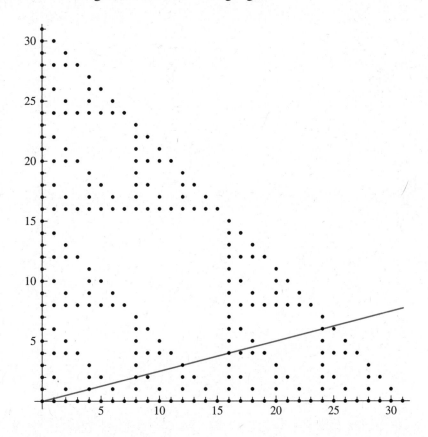

Notes

1. Wacław Franciszek Sierpiński was a Polish mathematician. He was known for contributions to set theory, number theory, theory of functions, and topology. He published over 700 papers and 50 books. Three well-known fractals are named after him, as are Sierpinski numbers and the associated Sierpinski problem.
2. Primes of the form $((2)^2)^m + 1$ are called Fermat primes.

CHAPTER 20

Nice Numbers

Learning Objectives

- Add reciprocals of positive integers to represent any rational number.
- Identify nice numbers.

Nice numbers involve working with some tricky arithmetic, understanding a layered mathematical definition, and constructing elegant solutions. This unit is accessible to high school students.

Definitions

Reciprocals. A reciprocal, or multiplicative inverse, is simply two numbers that have a product of 1. To find the reciprocal of a whole number, just turn it into a fraction in which the original number is the denominator and the numerator is 1.

Unit fraction. A fraction with a numerator of 1.

Sums of reciprocals of positive integers have been used to express other rational numbers since the ancient times. In particular, because the ancient Egyptians used them in such a way, the reciprocals of positive integers are referred to as Egyptian fractions. In 1202, Pisano showed that every rational number can be represented as a sum of distinct Egyptian fractions.

$$r = 1/a_1 + 1/a_2 + \ldots + 1/a_k, \text{ with } 1 \leq a_1 < a_2 < \ldots < a_k$$

In particular, every positive integer can be represented in this form. As follows, we consider representations of the unity:

$$1/a_1 + 1/a_2 + \ldots + 1/a_k = 1, 1 \leq a_1 < a_2 < \ldots < a_k$$

where the terms in the representations are not necessarily distinct. For example,

(1) $1/1 = 1$
(4) $1/2 + 1/2 = 1$
(6) $1/2 + 1/3 + 1/6 = 1$.

The sum of the denominators in the equations above are called "nice numbers."

Example 1

Is 9 a nice number?

Solution

Yes, $3 + 3 + 3 = 9$; $1/3 + 1/3 + 1/3 = 1$.

Example 2

Is 2 a nice number?

Solution
The only addends are 1 and 1; made into reciprocals, $1/1 + 1/1 \neq 1$.

Example 3

Are 3 and 5 nice numbers?

Solution
They are not; 5 is not a nice number because when we add its reciprocal addends, it doesn't equal 1.

By definition, then, an integer n will be called nice if it can be written as $n = a_1 + a_2 + \ldots + a_k$, where $a_1, a_2, \ldots, a_k$ are positive (not necessarily distinct) such that $1/a_1 + 1/a_2 + \ldots + 1/a_k = 1$.

Find the greatest common factors to reduce fractions, and use least common multiple to add fractions.

Show the algorithm for adding, subtracting, multiplying, and dividing fractions.

CHAPTER 20: NICE NUMBERS

PROBLEMS

1. Show that every perfect square is nice.
2. Observe that $a_1, a_2, \ldots, a_k$ are summands (terms) in a representation of a nice number, then $1/2a_1 + 1/2a_2 + \ldots + 1/2a_k = 1/2$.

   Show that if n is a nice number, then so is $2n + 2$.
3. Show that 10, 20, 22, 24, and 34 are nice numbers.

   Observe that $1/4 + 1/4 = 1/3 + 1/6 = 1/2$, and then solve the following exercises.
4. Show that 17 and 18 are nice numbers.
5. Show that if n is a nice number, then so are $2n + 8$ and $2n + 9$.
6. Use the previous result to show that 26, 27, 28, 29, 30, and 31 are all nice numbers.

SOLUTIONS

1. We know that $n^2 = \underbrace{n + n + \ldots + n}_{n \text{ times}}$. If $a_1 = n, a_2 = n, \ldots, a_n = n$, then

   $$\frac{1}{a_1} + \frac{1}{a_2} + \ldots + \frac{1}{a_k} = \underbrace{\frac{1}{n} + \frac{1}{n} + \ldots + \frac{1}{n}}_{n \text{ times}} = n \cdot \frac{1}{n} = 1$$

   and then n^2 is a nice number.

2. Let $n = a_1 + a_2 + \ldots + a_k$. From the condition we know that $1/a_1 + 1/a_2 + \ldots + 1/a_k = 1$. We know that $2a_1 + 2a_2 + \ldots + 2a_k + 2 = 2n + 2$. Let $b_1 = 2a_1, b_2 = 2a_2, \ldots, b_k = 2a_k, b_{k+1} = 2$. Then $1/b_1 + 1/b_2 + \ldots + 1/b_{k+1} = 1/2a_1 + 1/2a_2 + \ldots + 1/2a_k + 1/2 = 1/2 + 1/2 = 1$.

3. We know that $10 = 2 \times 4 + 2$. Therefore, to show that 10 is nice we need to show that 4 is nice. But we also know $4 = 2 \times 1 + 2$. Thus, to show that 4 is nice, we need to show that 1 is nice, which is obvious. We thus conclude that 10 is also nice.

   - $20 = 2 \times 9 + 2$, $9 = 3 + 3 + 3$, $1/3 + 1/3 + 1/3 = 1$; thus, 20 is nice.
   - $22 = 2 \times 10 + 2$, and we showed 10 is nice; therefore, 22 is also nice.
   - $24 = 2 \times 11 + 2$, $11 = 2 + 3 + 6$, $1/2 + 1/3 + 1/6 = 1$; therefore, 24 is also nice.
   - And for 34, $2 \times 16 + 2$, $16 = 4 + 4 + 4 + 4$, $1/4 + 1/4 + 1/4 + 1/4 = 1$, so 34 is nice, too.

   Also, $16 = 2 \times 7 + 2$, but 7 is not nice, and this route does not descend to a nice number, so we go to the other route to prove it is nice.

4. We know that $17 = 4 + 4 + 6 + 3$, and that $1/4 + 1/4 + 1/6 + 1/3 = 1/2 + 1/2 = 1$. This proves that both 17 and 18 are nice numbers.

5. Since $n = a_1 + a_2 + \ldots + a_k$ is a nice number, then we know that $2n + 2$ is also a nice number (see Exercise 2). Thus, if we use the same approach as in Exercise 2, we let $b_1 = 2a_1, b_2 = 2a_2, \ldots, b_k = 2a_k, b_{k+1} = 2$. Then

$$\frac{1}{b_1}+\frac{1}{b_2}+\ldots+\frac{1}{b_{k+1}}=\frac{1}{2a_1}+\frac{1}{2a_2}+\ldots+\frac{1}{2a_k}+\frac{1}{2}$$

$$=\frac{1}{2a_1}+\frac{1}{2a_2}+\ldots+\frac{1}{2a_k}+\frac{1}{4}+\frac{1}{4}$$

$$=\frac{1}{2a_1}+\frac{1}{2a_2}+\ldots+\frac{1}{2a_k}+\frac{1}{3}+\frac{1}{6}$$

$$=1$$

This goes to prove that both $2n+8$ and $2n+9$ are nice numbers.

$26 = 2 \times 9 + 8$, $9 = 3 + 3 + 3$, $1/3 + 1/3 + 1/3 = 1$; thus, 26 is nice.

$27 = 2 \times 9 + 9$; thus, 27 is nice.

$28 = 2 \times 10 + 8$, we showed before that 10 is nice; thus, 28 is also nice.

$29 = 2 \times 10 + 9$, we showed before that 10 is nice; thus, 29 is also nice.

$30 = 2 \times 11 + 8$, we showed before that 11 is nice; thus, 30 is also nice.

$31 = 2 \times 11 + 9$, we showed before that 11 is nice; thus, 31 is also nice.

Index

A

Academic capital, 113–114
Accessibility, 16
Accountability, 111
Active creativity, 54
Active engagement, 62, 68
Active learning, 59
Active learning environment, 91–97
Activities, 150
Additional practice, 166
ADHD. *See* Attention-deficit hyper-active disorder
Agile development, 35
AIME. *See* American Invitational Mathematics Exam
Aldrin, Buzz, 55
Algebraic concepts, 199, 245
Algebraic identities, 87–88
Algebraic inequalities, 89
Algebraic representations, 187
AMC. *See* American Mathematics Competitions
American Invitational Mathematics Exam (AIME), 103–104, 106
American Mathematics Competitions (AMC), 103–104, 105
American Regions Mathematics League (ARML), 106
Ancient number systems, 9
Angle bisectors, 91
Angles: definitions of, 177; parallel lines and, 177–183; in polygons, 178–179; in triangles, 180–183
Anxiety, 64
Apothem polygons, 237
Approaches: bottoms-up, 36–37; core values in, 43–44; discovery in, 38–41; fixed mindset, 27–28; flipped classrooms, 37–38; for individuals, 36–37; mission statements for, 41–43; sage on the stage approach for, 20; systems for, 35–36; in testing, 45; top-down, 36–37
Area, 91, 165, 172–173, 235–236, 239
Arithmetic progressions, 221, 223
ARML. *See* American Regions Mathematics League
Attention-deficit hyper-active disorder (ADHD), 67
Attitude, 61
Authenticity: cargo cult science and, 21–22; for students, 22–24; by teachers, 94
Autonomy, 47–48
Averages, 88
AwesomeMath Enrichment programs, 5, 41, 121–122; play in, 43; recitation in, 101; students in, 55; suggestion box for, 45, 49. *See also* Teacher inspiration

B

Bach, Johann Sebastian, 51
Beethoven, Ludwig van, 51
Behavior, 49, 93
Benjamin, Arthur, 155
Biographies, 18
Boards, 98–99
Bolyai, F., 239
Bolyai-Gerwien theorem, 239–243
Bolzano's theorem, 138
Bottoms-up approach, 36–37
Brocard-Ramanujan problems, 194

C

Calculators, 76
Campbell, Joseph, 38–39

Cantor, Georg, 48
Capital, 111, 113–114
Cargo cult science, 21–22
Carnegie Mellon University, xvii–xviii
Case studies: on AwesomeMath Enrichment programs, 5; on education, 2; on enrichment schools, 119; on geometry, 213; on information, 14; on math circles, 17; on MATHCOUNTS teams, 97; on research, 8; square roots in, 22; on teachers, 58; on teaching assistants, 80
Catalan's identity, 88
Cayley, Arthur, 160
Chalkboards, 98–99
Challenge: collaboration and, 42; in community outreach, 92–93; Conrad Challenge, 107; continuous, 60, 63, 68; discovery and, 21; engagement from, 15, 80; M3 Challenge, 105; Paul Erdös International Math Challenge, 104–105; Pythagorean Theorem for, 33; skill gaps as, 66; for students, 76–77, 150, 153–154, 162–163; from teachers, 10
Chemistry, 107
Chess, 170–171
Choices, 53–54
Chunking, 80–81
Circles, 219
Clarity, 42–43, 51
Classrooms: collaboration in, 112; environment for, 79–81; as flipped, 37–38, 58, 101; management of, 93–96
Coaching. See Teachers
Collaboration, xviii; challenge and, 42; in classrooms, 112; in community outreach, 6; competition and, 73; creativity from, 36–37, 79; in environment, 15; groups for, 10, 50; identity capital in, 113–114; innovation and, 76–77; leadership in, 108–112; learning from, 92; mathematical games for, 114–116; in PBL, 48, 55, 58, 112, 123–125, 130, 135; rewards from, 51; rotation and, 96; schedules for, 118–119; skills from, 3, 7; for students, 107–108; by teachers, 29; trade in, 113
College, 71, 126–127
Combinations, 248–249
Combinatorial set theory, 228
Combinatorics, 7, 191, 245
Communication, 98, 109
Community: competition and, 117–118; map of engagement for, 120–121; math circles for, 119; math clubs for, 118–119; parents in, 120–121; PBL and, 103; PLC, 17, 119–120
Community outreach: challenge in, 92–93; collaboration in, 6; experience in, 3; PBL in, 19, 92; teachers in, 40, 54, 112

Competition: collaboration and, 73; community and, 117–118; engagement and, 3; for groups, 15; Mandelbrot Competition, 105; in mathematics, 103–106; in PBL, 106–107; purpose with, 50–52; STEM and, 16; timing in, 78
Complex problems, 24
Computational linguistics, 7
Concentric circles, 219
Confidence, 98, 109
Congruent triangles, 168, 177
Connection, 42
Conrad Challenge, 107
Consecutive numbers, 185–189, 227
Continuous challenge, 60, 63, 68
Control, 65
Conversations, 92–93
Coopertition, 15, 50, 103
Coping, 64
Co-prime integers, 227–229
Core values, 43–44
Cornell method, 99–101
Counting, 20
Creativity: active, 54; from collaboration, 36–37, 79; flexibility and, 29, 73; innovation and, 109–110; introspection and, 31–32; play and, 96; research on, 77; from resilience, 10–12
Crisan, Vlad, 131–132
Critical thinking, xviii; pedagogy for, xix; about Pythagorean Theorem, 77; rigor for, 9–10; scale in, 20; understanding and, 24
Cryptarithmetic, 151–155
Cubes, 84
Curiosity: engagement and, 49; feedback and, 50; practice and, 99; from range, 5, 8; as relentless, 77–78
Curriculum: for education, xviii; incomplete, 61; lectures in, 38; for PBL, 14–15, 39–40, 52–53; relevance in, 18–21, 62; risk in, 1. See also Problem-based curriculum

D

da Vinci, Leonardo, 150
Decimals, 81–82
Decisiveness, 110
Deductive reasoning, 34
Deeper learning, 59, 62, 68
Definitions: of angles, 177; for area, 165; for consecutive numbers, 185; for dissection time, 239; for factorials, 191; for nice numbers, 255; for Pigeonhole principle, 227; for polygonal numbers, 205; for polygons, 166, 205; Pythagorean theorem, 213; for sequences, 221; of triangles, 177; for triangular numbers, 199; for Viviani's theorem, 235

Descartes, René, 144
Design thinking, 35
Diagonals, 202, 214–215, 218, 240
Diplomacy, 110
Dirichlet's box principle. *See* Pigeonhole principle
Discovery: challenge and, 21; engagement from, 18–19; from play, 12; risk and, 38–41; speed and, 47–48; teachers and, 14
Discrete math, 2–3, 69–70
Dissection time, 239–243
Diversity, 111–112
Divisibility, 86
Doctors, 22
Drive (Pink), 47
Dry erase boards, 98–99
Dudeney, H. E., 151

E

Eddington, Arthur Stanley, 146
Education: case studies on, 2; curriculum for, xviii; human-centric pedagogy for, 36; parents in, 57; PBL in, xviii
Educators. *See* Teachers
Efficacy: in communication, 98; in learning, 92–93, 123, 127, 129–136; in PBL, 124; of strategies, 23
Efficiency, 33, 61
Ego, 27–29
Einstein, Albert, 18
Embedding proofs, 237–238
Empathy, 64
Encouragement: from parents, 121; for students, 79–80; from teachers, 49
The End of Average (Rose), 36
Engagement: active, 62, 68; from challenge, 15, 80; competition and, 3; with complex problems, 24; curiosity and, 49; from discovery, 18–19; goals and, 53; with logic problems, 38; map of, 70–71, 120–121; with parents, 37; from play, 8, 17; with process, 121; strategies for, 54–55; with students, 75–76; by teachers, xviii
Enrichment centers, 25
Enrichment schools, 119
Environment: for classrooms, 79–81; collaboration in, 15. *See also* Learning environment
Equiangular polygons, 168–170
Equilateral polygons, 168–170, 236
Equilateral triangles, 241
Ergodic theory, 229, 233
Euler, Leonhard, 160, 216
Euler bricks, 213, 216–219
Evaluation, xvii
Even numbers, 185
Exercises, 9
Expectations, 27–28, 44

Experience: in community outreach, 3; inexperience, 63, 69; judgments and, 113; with PBL, 128–129, 132–134, 136–137; for teachers, 125–126, 136
Exploration, 9

F

Factorials, 85, 191–197
Fagnano, Giulio Carlo de' Toschi di, 145
Fear, Uncertainty, and Doubt (F.U.D.), 8
Feedback: curiosity and, 50; in PBL, 56; from teachers, 48
Fermat, Pierre de, 144–145
Fermat primes, 253
Feynman, Richard, 21–22
Fibonacci numbers, 18, 84–85, 225, 247
Field, Rachel, 54
Financial capital, 113–114
Finite sets, 157–159, 239
Fixed mindset approaches, 27–28
Flexibility, 29, 73
Flipped classrooms, 37–38, 58, 101
Flow, 31
Focus, 63
Forestry, 165
Fractions, 81–82, 255–258
Franklin, Benjamin, 160
F.U.D. *See* Fear, Uncertainty, and Doubt
Fun. *See* Play

G

Gains: from learning, 57; for parents, 68; for students, 62–63; for teachers, 59
Galois theory, 145
Game theory, 7, 11, 18
Games, 111, 114–116
Gaps, in skills, 60, 63, 66, 68, 70
Gardner, Howard, 30
Geometry, 245; angle bisectors in, 91; case studies on, 213; geometric progressions, 221–226; isosceles trapezoids, 90; knowledge in, 131; parallelograms, 89; quadrilaterals, 90; rectangles, 89; rhombi, 89; right triangles, 90; similar triangles, 90–91; squares, 90; Theorem of Thales in, 90; trapezoids, 90–91; triangles midlines, 90. *See also* Viviani's theorem
Gerwien, Klaus, 239
Gifted learners, 66, 119
Gladwell, Malcolm, 53
Goals: engagement and, 53; strategies for, 45; for students, 43–44
GPA. *See* Grades
Grace: in learning environment, 27, 29–30; under pressure, 98; skills from, 78–79

Grades, 69–71
Grid, 165
Groups: activities for, 150; clarity for, 51; for collaboration, 10, 50; competition for, 15; math circles for, 16–18; in PBL, 126; play for, 7
Growth mindset, 35
Gudder, S., 14
Guessing, 223
Guidance, by teachers, 56

H

Hardy, G. H., 6
Heron's formula, 219
Hero's Journey (Campbell), 38–39
Herzig, Emily, 6, 92, 124–125
Hexagonal numbers, 208
High School Mathematical Contest in Modeling (HiMCM), 105
Hook problems, 61
Hooke, Robert, 55–56
The Housekeeper and the Professor (Ogawa), 15
Hubris, 29
Human-centric pedagogy, 19, 36
Humility, 109

I

Idealism, 76
Identities: algebraic, 87–88; capital in, 111, 113–114; Catalan's, 88; Lagrange's, 88
Imagination, 30
IMO. *See* International Mathematics Olympiad
Impostor Syndrome, 39, 64
Inclusion, 111–112
Independence, 43
Individuals, 36–37
Inductive reasoning, 34
Inequalities, 89
Inexperience, 63, 69
Information: case studies on, 14; for students, 5; testing and, 9; understanding and, 13, 23
Innate talent, 99
Innovation: collaboration and, 76–77; creativity and, 109–110; rewards for, 42
Inquiry, 48–50
Inspiration: for learning environment, 44–46; from logic problems, 8. *See also* Teacher inspiration
Integers, 83–84, 185–189, 199; co-prime, 227–229; factorials and, 194; nonnegative integer exponents, 88; positive, 255–258; properties of, 213; sequences and, 245–246
Integrity: for learning, 21–22; principles and, 110–111
Intellectual risk, 39

Interaction, 91
International Mathematics Olympiad (IMO), 3, 80, 106
International Olympiad in Informatics (IOI), 107
International Physics Olympiad (IPhO), 3
Interpersonal skills, 59, 62, 68
Intervals, 227
Introduction to Arithmetic (Plutarchus), 201, 204
Introspection, 30–32
Intuition, 163
Invariance, 168
IOI. *See* International Olympiad in Informatics
IPhO. *See* International Physics Olympiad
Isosceles trapezoids, 90

J

Joyce, James, 39
Judgments: experience and, 113; patience in, 97–98; for students, 97; by teachers, 80
Jung, Carl, 113

K

Kelley, David, 35
Kindness, 43
Kisačanin, Branislav, 3, 40, 54, 134–135, 213
Knowledge: banks, 81–91; in geometry, 131; for PBL, 23; prior, 163; process for, 78–79; for students, 27–30, 44–45

L

Lagrange's Identity, 88
Langarica, Alicia Prieto, 112, 125–128
Large numbers, 191
Leadership, 96, 108–112
Learning, xviii; authenticity in, 21–24; from collaboration, 92; deeper, 59, 62, 68; efficacy in, 92–93, 123, 127, 129–136; efficiency in, 33; enrichment centers for, 25; expectations and, 27–28; facilitation of, 22; flow for, 31; gains from, 57; gifted learners, 66, 119; integrity for, 21–22; lesson-specific questions in, 97–98; metrics for, 48; with music, 7; outcome-based, 2; pains from, 57; from play, xviii–xix, 1; PLCs, 17, 119–120; process for, 41–42; relevance in, 13–21; student-centric, xviii, 37, 59, 62, 68; styles of, 30; talent in, 52–53; training and, 15–16; usefulness in, 25. *See also* Problem-based learning
Learning environment, 3; active, 91–97; ego in, 27–29; grace in, 27, 29–30; inspiration for, 44–46; teachers and, 33–35; Venn diagrams in, 30–33. *See also* Approaches
Learning objectives. *See* Mini-units

Lectures: in curriculum, 38; sage on the stage approach for, 20; by speakers, 18
Lesson plans. *See* Mini-units
Lesson-specific questions, 97–98
Lines, 177–183
Listening, 108–109
Local peers, 120
Logic problems, 7; engagement with, 38–39; inspiration from, 8; relevance in, 19; in suggestion box, 18
Lucas, Édouard, 160
Luck, 55–56

M

M3 Challenge. *See* MathWorks Math Modeling Challenge
Magic squares, 159–163
Management, 93–96
Mandelbrot Competition, 105
Manipulatives, 17, 50, 163
Manual of Harmonics (Plutarchus), 201, 204
Map of engagement, 70–71, 120–121
Mastery: purpose and, 47; research on, 53; by students, xvii–xviii; vulnerability in, 49
Math Olympiad for Elementary and Middle School (MOEMS), 104
Math Olympiad Program (MOP), 117
Mathematics: AIME, 103–104, 106; AMC, 103–104, 105; chess and, 170–171; competition in, 103–106; discrete math, 2–3, 69–70; games with, 111, 114–116; math circles, 16–18, 112, 119; math clubs, 118–119; *Math for Mathletes*, 213; math kangaroo, 105; MATHCOUNTS teams, 97, 104, 122; Mathematical relevance, 14–18; *Mathematical Snapshots*, 165; mental mathematics, 155–156; New math, 68; phobia of, 37, 79. *See also specific topics*
MathWorks Math Modeling (M3) Challenge, 105
Meaningful problems, 75–79
Means, 227
Meditation, 94
Mental mathematics, 155–156
Methods: Pigeonhole principle as, 228; for teachers, 123–125, 127, 130, 134; trial-and-error, 163
Metrics, for learning, 48
Metroplex Math Circle, 112
Midlines, 90
Mindsets: fixed, 27–28; growth, 35; what if, 78
Mingus, Charles, 10
Mini-units: cryptarithmetic, 151–155; elements of a finite set, 157–159; equilateral polygons, 168–170; magic squares, 159–163; math and chess, 170–171; for pains, 61; for PBL, 147; Pick's theorem, 165–168; polygons, 168–170; questions for, 147–148; roman numeral problems, 148–150; spheres, 172–173; squaring numbers, 155–156; toothpicks math, 163–165
Mission statements, 41–43
Mistakes: in PBL, 23–24; phobia of, 97; understanding, 97–99
Modeling, 49, 93, 121
MOEMS. *See* Math Olympiad for Elementary and Middle School
Monomyth (Campbell), 38–39
Monroe, Marilyn, 150
MOP. *See* Math Olympiad Program
Motivation, 111
Movement, 21
Mozart, Wolfgang Amadeus, 51
Multiple events, 87
Mushkarov, Oleg, 136–144
Music: learning with, 7; Mozart tables in, 51; Pythagorean Theorem in, 32

N

NACLO. *See* North American Computational Linguistics Olympiad
New math, 68
Newberry, Anthony, 122–124
Newton, Isaac, 55–56
Nice numbers, 255–258
Nonnegative integer exponents, 88
North American Computational Linguistics Olympiad (NACLO), 106
Notations, 157
Note taking, 99–101
Numbers: ancient systems for, 9; consecutive, 185–189, 227; even, 185; Fibonacci, 18, 247; hexagonal, 208; large, 191; nice, 255–258; odd, 185, 218; polygonal, 205–211; positive real, 88; prime, 82–83, 201, 227, 253; Sierpiński, 253; square, 206, 209; squaring, 155–156; tetrahedral, 203, 249; theory for, 245, 253; triangular, 199–204, 206–210, 246
Nurturing, of talent, 99

O

Observation, 149
Octagons, 241–242
Odd numbers, 185, 218
Ogawa, Yoko, 15
Online peers, 120
Outcome-based learning, 2
Outcomes, 70
Outliers (Gladwell), 53

P

Pacing, 37, 60
Pains: grades as, 70–71; from learning, 57; mini-units for, 61; for parents, 68–69; for students, 63–64; for teachers, 60–61. *See also* Problem-based curriculum
Parallel lines, 177–183
Parallelograms, 89, 91
Parents: in community, 120–121; discrete math for, 69–70; in education, 57; engagement with, 37; gains for, 68; grades for, 69; modeling by, 121; pains for, 68–69; problem-based curriculum for, 67–71; schedules for, 69; teachers and, 70; value propositions for, 67
Pascal's triangle, 86, 245–253
Patience, 97–98, 110
Pattern recognition, 185
Pauk, Walter, 99–101
Paul Erdös International Math Challenge, 104–105
Pausch, Randy, xvii–xviii
Pedagogy: for critical thinking, xix; of exercises, 9; human-centric, 19, 36; for PBL, 1–3
Peers, 120
Peer-tutoring, 95
People, 54–55
Percentages, 81–82
Perfect boxes, 216
Perfectionism, 64
Permutations, 87
Perpendicular lines, 177
Perspective, 59, 62, 67, 111
Phobia, 37, 79, 97
Physical capital, 113–114
Physics, 7, 107
Pick, Georg, 165–168
Pick's theorem, 165–168
Pigeonhole principle, 86–87, 227–233
Pink, Daniel, 47
Planning: for students, 11–12; by teachers, 44–46
Play: in AwesomeMath Enrichment programs, 43; creativity and, 96; discovery from, 12; engagement from, 8, 17; for groups, 7; imagination and, 30; learning from, xviii–xix, 1; manipulatives for, 50; for PBL, 29; roleplaying, 49; work compared to, 6
PLCs. *See* Professional learning communities
Plutarchus, Lucius Mestrius, 201, 204
Polygonal numbers, 205–211
Polygonal surfaces, 239
Polygons, 168–170, 236–238; angles in, 178–179; definition of, 166, 205; sequences with, 248; squares and, 240
Pompe, Waldemar, 136
Positive integers, 255–258
Positive real numbers, 88
Post reflections, 144–146
Practice: additional, 166; curiosity and, 99; for receiving, 108
Praise, 95
Presentation, 98–99
Prime numbers, 82–83, 201, 227, 253
Principles: of inclusion-exclusion, 20; integrity and, 110–111; pigeonhole, 86–87, 227–233
Prior knowledge, 163
Problem-based curriculum: for parents, 67–71; for students, 61–67; systems for, 57; teachers, 58–61
Problem-based learning (PBL), xvii; autonomy in, 47–48; collaboration in, 48, 55, 58, 112, 123–125, 130, 135; community and, 103; in community outreach, 19, 92–93; competition in, 106–107; curriculum for, 14–15, 39–40, 52–53; in education, xviii; efficacy in, 124; expectations in, 44; experience with, 128–129, 132–134, 136–137; feedback in, 56; full units for, 175; games in, 114–115; groups in, 126; knowledge for, 23; meaningful problems in, 75–79; mini-units for, 147; mistakes in, 23–24; pedagogy for, 1–3; play for, 29; principle of inclusion-exclusion for, 20; reasoning in, 33–34; relevance in, 77; rewards for, 5–12; scale in, xviii; schedules and, 63, 66; speed in, 13; suggestion box for, 7; for teachers, 73–74; in think tank, 39; understanding in, 113–114. *See also* Resources
Process: engagement with, 121; for knowledge, 78–79; for learning, 41–42; outcomes and, 70
Professional learning communities (PLCs), 17, 119–120
Progressions, 221–226
Proofs. *See specific topics*
Purple Comet! Math Meet, 80–81, 104
Purpose: with competition, 50–52; decisiveness and, 110; mastery and, 47; research on, 47
Puzzles: game theory and, 18; for intuition, 163; in suggestion box, 19
Pythagorean Theorem, 32–33, 77, 213–219, 239–240
Pythagoreanism, 200, 204

Q

Quadrants, of success, 52–56
Quadrilaterals, 90
Questions: in Cornell method, 100–101; lesson-specific, 97–98; for mini-units, 147–148; for teachers, 77

R

Range: curiosity from, 5, 8; exploration and, 9; rigor and, 11
Ratios, 191
Reasoning, 33–34

Receiving (listening), 108
Reciprocals, 255–258
Recognition systems, 108
Rectangles, 89, 213–219, 240
Reflective thinking, 14, 147–148; post reflections, 144–146; resources for, 99–101
Regular polygons, 168
Relativity, 147–148
Relentless curiosity, 77–78
Relevance: in curriculum, 18–21, 62; in logic problems, 19; mathematical, 14–18; in PBL, 77; for students, 13–14
Research: case studies on, 8; in college, 126–127; on creativity, 77; on mastery, 53; on problem solving, 2; on purpose, 47; on STEM, 103–104
Resilience, 10–12
Resources: for active learning environments, 91–97; classroom environment and, 79–81; knowledge banks, 81–91; mistakes as, 97–99; for reflective thinking, 99–101; for talent, 99
Retention: in college, 71; Cornell method for, 99–101; for parents, 68; of students, 62
Revisions, 147–148
Rewards: from collaboration, 51; for innovation, 42; for PBL, 5–12
Rhombi, 89, 218
Right angles, 177
Right triangles, 90
Rigor, 9–11
Risk, 1, 38–41
Robinson, Ken, 77
Roleplaying, 49
Roles, for teachers, 108
Roman numeral problems, 148–150
Rose, L. Todd, 36
Rotation, 96, 108
Rule of divisibility, 86

S

Sage on the stage approach, 20, 37
Sample problems, 9–10
Scaffolding, 9
Scale: accessibility and, 16; in critical thinking, 20; in material, 66; in PBL, xviii; for students, 17
Schedules: for collaboration, 118–119; for parents, 69; PBL and, 63, 66; for students, 1
Schlicter, Dean, 18
Science, 7, 107
Science, technology, engineering, mathematics (STEM), 3; competition and, 16; discrete math in, 70; math circles for, 119; peers in, 120; research on, 103–104
Scores, 63, 68
Self-assessment, 23

Sendova, Jenny, 137–146
Sensory-seeking, 66
Sequences, 205, 221–226, 245–246, 248
The Shape of Space (Stottile), 17
Sierpiński, Franciszek, 253
Sierpiński numbers, 253
Similar triangles, 90–91, 177, 179
Skills: from collaboration, 3, 7; gaps in, 60, 63, 66, 68, 70; from grace, 78–79; interpersonal, 59, 62; for students, 58; testing and, 36
Social awareness, 64
Social capital, 113–114
Some People (Field), 54
Speakers, 18
Speed, 13, 47–48
Spheres, 172–173
Square numbers, 206, 209
Square roots, 22
Squares, 83, 90, 213–214, 229–231, 240
Squaring numbers, 155–156
Stakeholders. *See* Students
Standardized testing, 60, 63, 69
Stankova, Zvezdelina, 16
Steinhaus, H., 165
STEM. *See* Science, technology, engineering, mathematics
Stottile, Frank, 17
Straight angles, 177
Strategies: efficacy of, 23; for engagement, 54–55; for goals, 45; Pigeonhole principle as, 228; for solving, 23
Struggle, 13, 98–99
Students: active creativity for, 54; with ADHD, 67; attitude for, 61; authenticity for, 22–24; in AwesomeMath Enrichment programs, 55; challenge for, 76–77, 150, 153–154, 162–163; choices for, 53–54; collaboration for, 107–108; confidence for, 98; control for, 65; conversations between, 92; discrete math for, 2–3; dissection time for, 239; encouragement for, 79–80; in *The End of Average*, 36; engagement with, 75–76; evaluation of, xvii; factorials for, 191–197; gains for, 62–63; goals for, 43–44; guessing by, 223; Impostor Syndrome for, 39, 64; individual, 36–37; information for, 5; inquiry by, 48–50; interaction between, 91; introspection by, 30–31; judgments for, 97; knowledge for, 27–30, 44–45; mastery by, xvii–xviii; movement for, 21; nice numbers for, 255; numbers for, 185; observation by, 149; pains for, 63–64; Pascal's triangle for, 245; peer-tutoring for, 95; Pigeonhole principle for, 227; planning for, 11–12; polygonal numbers for, 205; praise for, 95; problem-based curriculum for, 61–67; Pythagorean Theorem for, 213; relevance for, 13–14; retention of, 62; roleplaying by, 49; scale

for, 17; schedules for, 1; self-assessment by, 23; sequences for, 221; skill gaps for, 70; skills for, 58; speakers for, 18; struggle for, 13, 98–99; student-centric learning, xviii, 37, 59, 62, 68; suggestion box for, 163; teachers and, 2, 92, 194; triangular numbers for, 199; upfront work for, 37; value propositions for, 62; Viviani's theorem for, 235; waiting for, 6. *See also* Groups

Styles, 30, 45–46

Subsets, 227

Success, 52–56

Suggestion box: ancient number systems in, 9; for anxiety, 64; for AwesomeMath Enrichment programs, 45, 49; biographies in, 18; game theory in, 11; logic problems in, 18; movement in, 21; for PBL, 7; puzzles in, 19; for students, 163; for teachers, 51

Supplementary angles, 177

Support, 69

Suspense, 50

Systems: for approaches, 35–36; for problem-based curriculum, 57; for recognition, 108; for understanding, 45–46

T

Talent, 52–53; innate, 99; nurturing of, 99; resources for, 99

Teacher inspiration: AwesomeMath Enrichment programs and, 121–122; from Crisan, 131–132; from Herzig, 124–125; from Kisacanin, 134–135; from Langarica, 125–128; from Mushkarov, 136–144; from Newberry, 122–124; from Pompe, 136; post reflections after, 144–146; from Turcas, 132–134; from Yotov, 128–131

Teachers: assistants, 80; authenticity by, 94; case studies on, 58; challenge from, 10; chunking by, 80–81; classroom management for, 93–96; collaboration by, 29; in community outreach, 40, 54, 112; confidence for, 109; coping and, 64; discovery and, 14; doctors and, 22; efficiency for, 61; encouragement from, 49; engagement by, xviii; experience for, 125–126, 136; feedback from, 48; gains for, 59; guidance by, 56; *The Housekeeper and the Professor*, 15; humility for, 109; idealism for, 76; inexperience of, 63, 69; judgments by, 80; learning environment and, 33–35; methods for, 123–125, 127, 130, 134; pains for, 60–61; parents and, 70; PBL for, 73–74; planning by, 44–46; PLCs for, 17; problem-based curriculum, 58–61; questions for, 77; roles for, 108; students and, 2, 92, 194; styles for, 45–46; suggestion box for, 51; value propositions for, 59. *See also* Resources

Teamwork, 95

Teenagers, 113

Teleology, 47–56

Testing: approaches in, 45; information and, 9; scores in, 63, 68; skills and, 36; standardized, 60, 63, 69

Tetrahedral numbers, 203, 249

Theorems: Bolyai-Gerwien theorem, 239–243; Bolzano's theorem, 138; Ergodic theory, 229, 233; Galois theory, 145; Pick's theorem, 165–168; Pythagorean Theorem, 32–33, 77, 213–219, 239–240; Theorem of Thales, 90; theory, 191, 228, 245, 253; Viviani's theorem, 168, 235–238. *See also* Game theory

Think tank: counting problems in, 20; PBL in, 39; scaffolding in, 9

Three C's. *See* Collaboration; Community; Competition

Time, 239–243

Timing, 55–56, 60, 78

Toothpicks math, 163–165

Top-down approach, 36–37

Torricelli, Evangelista, 145

Trade, 113

Training, 15–16

Transformation, 239

Trapezoids, 90–91

Trial-and-error, 163

Triangles: angles in, 180–183; area of, 235–236; combinations with, 248–249; congruent, 168; definitions of, 177; equilateral, 241; Heron's formula for, 219; midlines of, 90; Pascal's triangle, 86, 245–253; prime numbers and, 201; similarity in, 90–91; triangular numbers, 199–204, 206–210, 246; triangulation, 239, 241. *See also* Polygonal numbers; Pythagorean Theorem; specific triangles

Trust, 107–108

Turcas, George Catalin, 132–134

U

Understanding: checks for, 150; critical thinking and, 24; information and, 13, 23; mistakes, 97–99; in PBL, 113–114; systems for, 45–46

Unit fractions, 255–258

Upfront work, 37

US National Chemistry Olympiad (USNCO), 107

USA Junior Mathematics Olympiad (USAJMO), 106

USA Mathematical Talent Search (USAMTS), 106

USA Mathematics Olympiad (USAMO), 106, 117

USA Physics Olympiad, 107

USAJMO. *See* USA Junior Mathematics Olympiad

USAMO. *See* USA Mathematics Olympiad
USAMTS. *See* USA Mathematical Talent Search
Usefulness, in learning, 25
USNCO. *See* US National Chemistry Olympiad

V

Value propositions: for parents, 67; for students, 62; for teachers, 59
Venn diagrams, 20, 30–33
Vertices, 202, 205–206
Vision, 111
Viviani, Vincenzo, 239
Viviani's theorem, 168, 235–238
Volume, 172–173
Vulnerability, 49

W

Waiting, for students, 6
What if mindsets, 78
Wilson prime, 191, 193–194
Work, 6, 37
World War II, 21

Y

Yotov, Mirroslav, 19, 128–131
YouTube, 145–146

Z

Zeros, 196–197